NCSM
Great Tasks for Mathematics 6-12

Engaging Activities for Effective Instruction and
Assessment that Integrate the Content and Practices
of the Common Core State Standards for Mathematics

Connie Schrock, Kit Norris,
David K Pugalee, Richard Seitz,
and Fred Hollingshead

Important notice regarding book materials

The National Council of Supervisors of Mathematics (NCSM) makes no warranty, either express or implied, including but not limited to any implied warranties of merchantability and fitness for a particular purpose, regarding any programs or book materials and makes such materials available solely on an "as-is" basis. In no event shall NCSM be liable to anyone for special, collateral, incidental, or consequential damages in connection with or arising out of the purchase or use of these materials, and the sole and exclusive liability of NCSM, regardless of the form of action, shall not exceed the purchase price of this product. Moreover, NCSM shall not be liable for any claim of any kind whatsoever against the use of these materials by any other party.

Permission is hereby granted to teachers to reprint or photocopy in classroom, workshop, or seminar quantities the pages in this book. These pages are designed to be reproduced by teachers for use in their classes, workshops or seminars. Such copies retain the NCSM copyright and may not be sold, and further distribution is expressly prohibited. Except as authorized above, prior written permission must be obtained from NCSM to reproduce or transmit this work or portions thereof in any other form or by any other electronic or mechanical means, including any information storage or retrieval system, unless expressly permitted by federal copyright law.

Copyright © 2013 The National Council of Supervisors of Mathematics. Except for the specific rights granted herein, all rights are reserved.

Printed in the United States of America.

ISBN: 978-0-9890765-1-7

Table of Contents

About this Book 3

Introduction 7

Acknowledgements 19

Grade 6 - Robot Arena 21

Grade 6 - The Missing Words 29

Grade 6 - How Many Fit? 39

Grade 7 - Sizing Up the Garage 47

Grade 7 - Most Square? 53

Grade 7 - Odd or Even 57

Grade 8 - How Do You Visualize It? 63

Grade 8 - The Shrinking Square 67

Grade 8 - Trouble with Trees 71

Number and Quantity - Irrational Thinking, Inc. 77

Number and Quantity - Stadium Dimensions 81

Algebra - Chilling Out 85

Algebra - Proving Patterns 89

Algebra - Waiting in the Queue 95

Algebra - Fractional Workers 101

Algebra - Dead Pennies 105

Functions - Ferris Wheel Gala 111

Functions - Rise and Shine 115

Functions - A Quadratic Smile 121

Functions - Trout Pond 127

Functions - Fractal Doodles 131

Geometry - The Tipi 137

Geometry - Infinity Pizza 141

Geometry - Quadrilaterals Flying High 145

Statistics and Probability - Classy Carnival 149

Statistics and Probability - Coaches' Dilemma 153

Statistics and Probability - How Fast Does it Fall? 157

Statistics and Probability - Searching Similarities 163

Statistics and Probability - YouTube™ Views 169

Appendix A: TI-Nspire™ - Plotting and Analyzing Data 173

Appendix B: TI-84 Plus - Plotting and Analyzing Data 181

About this Book

In 1996, the National Council of Supervisors of Mathematics officially released *Great Tasks and More!!* – the second in a series of timely NCSM Source Books for leaders in mathematics education. That volume provided an extensive set of camera-ready resources on the topic of mathematics assessment including performance tasks, sample rubrics, model transparencies and background articles. This first collection of great tasks was well received and NCSM members asked for a revision that supported and correlated with the Common Core State Standards for Mathematics (CCSSM, 2011).

In the summer of 2010, the NCSM Board charged a national writing committee to develop a set of Great Tasks for mathematics leaders and teachers that supported both the standards for mathematics content and the standards for mathematical practice as established by the CCSSM. The new standards set high expectations for mathematical learning for all students and for teaching. The initial question that the committee addressed was, "How can a society improve the mathematical proficiency of its students?" The answer was one simple measure – the degree to which educated students are able to take on the challenges of the society they inherit. That is, how will mathematical skills and knowledge help make the world a better place? Reflecting on student performance permeated the initial discussions. The first sample great task with teacher notes, activity launch, student work, and rubric was distributed at the spring 2011 annual NCSM meeting in Indianapolis, IN.

NCSM members applauded the early work of the committee and asked for additional and expanded great tasks. In the NCSM summer 2011 strategic plan, the NCSM Board developed a strand - Leadership Learning Resources - that commissioned the writing team to develop a series of K-12 Great Tasks that would help mathematics leaders and teachers assess student work. The second set of sample tasks were distributed at the spring 2012 annual NCSM meeting in Philadelphia, PA.

The purpose of this NCSM Great Tasks document-*NCSM Great Tasks for Mathematics 6-12: Engaging Activities for Effective Instruction and Assessment that Integrate the Content and Practices of the Common Core State Standards for Mathematics* is to provide 6-12 mathematics educators with a set of Great Tasks that can be used to engineer learning experiences to guide student to become problems solvers of the future. These Great Tasks will allow students to become inventors, to demonstrate capacity to think, to test opportunities to evaluate reasoning and examine sense-making. The Great Tasks will give teachers insight into student thinking through student-based evidence. The Board is extremely grateful to the writing team for their dedication, time, energy, and effort to research and write the great tasks. This book moved from an idea to reality because of the writers' wisdom, experience, and insight.

Ultimately, the ***NCSM Great Tasks for Mathematics 6-12: Engaging Activities for Effective Instruction and Assessment that Integrate the Content and Practices of the Common Core State Standards for Mathematics*** is designed to provide clear guidance on how to raise achievement in mathematics for *every student* and effectively implement the Common Core State Standards in Mathematics in *every classroom.* Because the great tasks in this collection were developed to illustrate both the philosophy and implementation needed to integrate great tasks in the curriculum, the tasks can also serve as models for professional development. Presenting these tasks in a workshop can involve discussing the sample tasks and supporting teacher materials, and then developing similar sets of tasks. Guiding questions for mathematics leaders in professional development might include:

- Is there evidence in the student work that demonstrates the use of the Standards for Mathematical Practice?
- How did the task expose a student's thinking about the content?
- What might a teacher plan as next steps as a result of the task?
- As leaders, what ways can these tasks be used with teachers and what is the intended outcome?

This book is the second part of a two-part series. The first book in the series is titled, ***NCSM Great Tasks for Mathematics K-5: Engaging Activities for Effective Instruction and Assessment that Integrate the Content and Practices of the Common Core State Standards for Mathematics.*** It is intended to provide K-5 mathematics educators with a set of Great Tasks that can be used to engineer learning experiences to guide secondary students to become problems solvers of the future. The NCSM Board hopes you find this compilation of Great Tasks interesting and helpful as a support structure that helps every student improve mathematically. The student work in the Great Tasks is a reminder that students do not always think and reason about mathematical ideas the way adults do. Rather the student work illustrates the need for teachers to rethink how students are engaged in the learning process and in deep understanding of mathematical ideas.

--Suzanne Mitchell, President of NCSM (2011-2013)

About the National Council of Supervisors of Mathematics and Its Members

The National Council of Supervisors of Mathematics (NCSM) is an international leadership organization for those who serve the NCSM vision of excellence and equity for student achievement in mathematics. NCSM is founded on the strength and dedication of a growing membership of mathematics education leaders. These leaders include grade-level team leaders, course-level team leaders, department chairs, district or county coaches and coordinators, site-based teacher leaders, district or provincial curriculum directors, principals, superintendents, college faculty and trainers of teacher leaders, and all who work to ensure the success of every child in mathematics.

NCSM was created at the 1968 Philadelphia meeting of the National Council of Teachers of Mathematics (NCTM) when a group of urban district supervisors decided that at the next annual meeting (Minneapolis, 1969) school district leaders should gather to form the National Council of Supervisors of Mathematics to address leadership issues in mathematics. An early and critical issue for NCSM was defining the membership. The founding members chose *not* to restrict membership to supervisors and instead welcomed all leaders and teachers of mathematics. The open membership theme has continued throughout the years; the 35 leaders who attended that first meeting in Minneapolis grew to more than 3,000 by the end of the 20th century.

As NCSM celebrates its 45th anniversary in 2013, the vision and ideals of our founders endure:

> N: Network and collaborate with stakeholders in education, business, and government communities to ensure the growth and development of mathematics education leaders.
>
> C: Communicate current and relevant research to mathematics leaders, and provide up-to-date information on issues, trends, programs, policies, best practices, and technology in mathematics education.
>
> S: Support and sustain improved student achievement through the development of leadership skills and relationships among current and future mathematics leaders.
>
> M: Motivate mathematics leaders to maintain a lifelong commitment to provide equity and access for all learners.

As NCSM enters its fifth decade, we continue to strive for excellence and equity for all children. Our greatest challenge is *developing the leadership knowledge and skills* that will advance these central vision points. To that end, the NCSM Board commissioned a

national writing team and enlisted the feedback of numerous professionals interested in mathematics education in order to create a two-part series of Great Tasks that mathematics educators can use to engineer learning experiences to guide student to become problems solvers of the future.
- *NCSM Great Tasks for Mathematics K-5: Engaging Activities for Effective Instruction and Assessment that Integrate the Content and Practices of the Common Core State Standards for Mathematics* and
- *NCSM Great Tasks for Mathematics 6-12: Engaging Activities for Effective Instruction and Assessment that Integrate the Content and Practices of the Common Core State Standards for Mathematics*

The NCSM Board establishes goals and creates and reviews new projects and initiatives that advance the mission and vision of the organization. The Board monitors the achievement of existing goals and projects to ensure continued alignment that supports the needs of mathematics education leaders. The NCSM Board invites you to become a part of a larger organization that provides professional learning opportunities necessary to support and sustain improved student achievement.

--The 2011-2012 and the 2012-2013 NCSM Board

Introduction

The Common Core State Standards for Mathematics Content

The education system of the United States has started down a clear path toward building and implementing a set of Common Core Standards for Mathematics Content with which all students in mathematics education grades K–12 will interact. These standards establish a set of high expectations for the mathematical learning of all students. They also provide guidance for parents, students, and educators to achieve these goals.

One way to help educators guide students in building proficiency at these practices is to model tasks and lessons like those presented in this publication.

What is a Great Task?

A Great Task:

- Revolves around an interesting problem – offering several methods of solution.
- Is directed at essential mathematical content as specified in the standards.
- Requires examination and perseverance – challenging students.
- Begs for discussion – offering rich discourse on the mathematics involved.
- Builds student understanding – following a clear set of learning expectations.
- Warrants a summary look back – with reflection and extension opportunities.

We know that students who innovate, create, discuss, engage, and are motivated to become problem solvers are more effective users of mathematics. Accordingly, mathematics instruction must focus on the development of these skills.

Presenting Great Tasks

"NCSM Great Tasks for Mathematics" are designed as follows. Each task contains a set of Teacher Notes that provide an overview of the task, the Common Core State Standards for Mathematical Content and for Mathematical Practice that the task requires, prerequisite understandings, and specific suggestions for using the task. Each task includes the following parts:

    1) an Activity Introduction that addresses key prerequisite understandings and assesses student readiness for the task;

    2) the Core Tasks which students are expected to perform individually and collaboratively; and

    3) a suggested Extension Activity to expand upon the learning within the Core Task.

In addition, a generic rubric is provided (see Figure 1) to help focus on the key aspects of completing a task.

What If ... ?

What if every classroom in America had students engaged in...

- Solving problems that took sense making and perseverance?
- Reasoning with number sense and extending ideas to abstract thinking?
- Defending and proving their ideas about mathematics?
- Constructing mathematical models to explore and solve real life problems?
- Using rulers, compasses, manipulatives, calculators, tablets, and computers in powerful ways?
- Using precise vocabulary and clearly-defined measurements to describe situations?
- Recognizing patterns, connections, and algorithms with an eye toward how they fit into the structure and logic of mathematics?

The answer is really quite simple. You would have classrooms that were meeting the promise of the Common Core State Standards in Mathematics. You would have classrooms educating citizens who would be up to the challenges of solving the problems the next generation will have to face!

The Practices in the Common Core State Standards

When the United States took on the task of building and implementing a set of Common Core Standards for all its students in mathematics education grades K–12, it also included a set of teaching and learning practices that set high expectations for mathematical learning for all students. The importance of these practices justifies their inclusion at the beginning of the documents for the Common Core Standards.

These practices provide a clear picture of what students will be **doing** as they construct their mathematical thinking. Classrooms will feature students engaging in learning and working as a community to ensure that everyone achieves.

Achieving a classroom that reflects the Standards for Mathematical Practice is no small feat. Careful planning with colleagues to define what students will be doing at every phase of the class period represents a vital first step towards incorporating these practices. Student engagement is essential to and a prerequisite for learning.

Standards for Mathematical Practice

"The Standards for Mathematical Practice describe varieties of expertise that mathematics educators at all levels should seek to develop in their students. These practices rest on important "processes and proficiencies" with longstanding importance in mathematics education. The first of these are the NCTM process standards of problem solving, reasoning and proof, communication, representation, and connections. The second are the strands of mathematical proficiency specified in the National Research Council's report Adding It Up: adaptive reasoning, strategic competence, conceptual understanding (comprehension of mathematical concepts, operations and relations), procedural fluency (skill in carrying out procedures flexibly, accurately, efficiently and appropriately), and productive disposition (habitual inclination to see mathematics as sensible, useful, and worthwhile, coupled with a belief in diligence and one's own efficacy)." See page 6 of http://www.corestandards.org/assets/CCSSI_Math%20Standards.pdf

As important as the Common Core Standards are, it is the Standards for Mathematical Practice that address thinking and acting like a mathematician and that call for students to regularly engage in the following:

1. Make sense of problems and persevere in solving them.
2. Reason abstractly and quantitatively.
3. Construct viable arguments and critique the reasoning of others.
4. Model with mathematics.
5. Use appropriate tools strategically.
6. Attend to precision.
7. Look for and make use of structure.
8. Look for and express regularity in repeated reasoning.

Knowing that we want students to think and perform in the manner described above opens us up to teaching practices that will provide students with learning opportunities that will help them develop into mathematical proficient students. Let us look at some of those teaching practices.

Teaching Practices to Support the Mathematical Practices

What teaching practices will support a strong implementation of the Common Core Practices? From research we know that effective teachers:

- Engineer lessons to build mathematical proficiency.
- Use formative assessment to understand what students know and identify action steps to enhance further learning.
- Foster student collaboration and opportunities for dialogue.
- Ask effective questions.

Engineer Lessons to Build Mathematical Proficiency

In order for students to become really proficient with a mathematical topic, they need to understand the concepts, develop fluency with any of the procedures that are involved, use several strategies for working with the ideas, communicate their thinking, and understand the importance or utility of what they are learning.

Good lessons are built to:

- Challenge and enrich student thinking by presenting meaningful tasks;
- Enhance understanding of the concepts;
- Develop mathematical vocabulary related to the ideas;
- Promote students' ability to learn several ways of looking at the ideas (graphical, numerical, pictorial, and algebraic);
- Develop fluency with procedures;
- Encourage students to use written or oral skills to explain what they know and to develop mathematical arguments;
- Enable students to deepen their understanding of concepts and make connections among various topics.

The lessons contained in **Great Tasks** have been designed to present meaningful tasks that will help students to deepen their thinking, learn to persevere, make connections, and communicate their understanding. Ultimately, students work to make the mathematics make sense.

The Great Tasks are designed to target content standards from the Common Core State Standards and feature several mathematical practices. As students work on these tasks, some may struggle, and this is appropriate and important for them to do so. Allow time for students to try to make sense of the context, try an approach, and then reflect on that approach. Support students by encouraging them to work with partners and share their thoughts after an initial independent period.

Use Assessments to Guide Teaching and Learning

The research is clear that teachers who use ongoing assessments to keep them abreast of what students understand are able to help clear up misconceptions and actually improve the pace of learning. Assessments can be short, in-class check-ins, warm-ups, exit tickets, quizzes, or unit tests. As long as the assessments are used to inform instruction and lead to action steps taken by students, the assessment serves to guide teaching and learning.

The Standards for Mathematical Practice focus on the active engagement of students throughout the class period. As teachers plan lessons, they need to reflect on exactly what the students are doing in each phase of the lesson. If students are sitting and watching the teacher or other students demonstrate a problem, the majority of the class is not engaged.

Foster Students' Collaboration and Opportunities for Dialogue

Here is a short set of suggestions for improving the level of students' classroom communications skills:

1. **Jot, Pair, Share:** After posing a question, provide time for students to think independently and jot down on paper their thoughts about the question. After a few moments, ask students to share what they had thought about with one other student. Students then discuss their thinking further.

2. **Let The Students Talk:** If you want to know the formula for the area of a triangle, ask the kids. Odds are that someone has already learned that formula. And when you get the answer $A = \frac{b \cdot h}{2}$, ask if someone can demonstrate why it works, ask what other formulas they know, ask them to come up with a problem that you could solve with the formula, and ask them to come up with one you couldn't solve.

3. **One Question is Seldom Enough:** Just as you get an answer might be the best time to ask if others agree. If they all agree, then ask who can prove that the answer makes sense. Knowing something is true is seldom as good as explaining how you know it is true. We all hear unsubstantiated claims in today's society, and our students really need to learn to support their reason with facts and logic!

4. **Wait Time is Great Time:** Giving time to pause before you allow people to answer a question shows the importance of the thinking that is essential for learning. The quick answers should be reserved for memorized facts, not for questions of substance. Even procedural tasks such as solving the equation $x^2 - 25 = 0$ might have lots of powerful mathematics that can be discussed if we give students time to ponder and think about ways to solve it. If you understand anything about teenagers, you quickly realize that they may make better choices if you get them to think before they leap!

Effective Questioning

Teachers can promote discourse and stimulate student thinking, and thereby develop the habits of mind suggested by the Standards for Mathematical Practice. Here is a list of questions from the *Professional Standards in Teaching Mathematics,* grouped into categories that reflect the mathematical practices.

* **Helping students work together to make sense of mathematics:**

  - "What do others think about what Janine said?"
  - "Do you agree? Disagree?"
  - "Does anyone have the same answer but a different way to explain it?"
  - "Would you ask the rest of the class that question?"
  - "Do you understand what they are saying?"
  - "Can you convince the rest of us that that makes sense?"

* **Helping students to rely more on themselves to determine whether something is mathematically correct:**

  - "Why do you think that?"
  - "Why is that true?"
  - "How did you reach that conclusion?"
  - "Does that make sense?"
  - "Can you make a model to show that?"

* **Helping students learn to reason mathematically:**

  - "Does that always work?"
  - "Is that true for all cases?"
  - "Can you think of a counterexample?"
  - "How could you prove that?"
  - "What assumptions are you making?"

* Helping students learn to conjecture, invent, and solve problems:
    - "What would happen if . . .? What if not?"
    - "Do you see a pattern?"
    - "What are some possibilities here?"
    - "Can you predict the next one? What about the last one?"
    - "How did you think about the problem?"
    - "What decision do you think he should make?"
    - "What is alike and what is different about your method of solution and hers?"

* Helping students to connect mathematics, its ideas, and its applications:
    - "What is the process for solving these types of problems?"
    - "How is this process like others you have used?"
    - "How does this relate to ____?"
    - "What ideas that we have already learned were useful in solving this problem?"
    - "Have we ever solved a problem like this one before?"
    - "What uses of mathematics did you find in the newspaper last night?"
    - "Can you give me an example of ____?"

What Does All This Have to Do with Great Tasks?

Turn the pages and see. We think you will find that it has everything to do with thinking about teaching and learning mathematics!

Using Great Tasks

There are many ways we believe this collection can be used. Tasks can be incorporated into mathematics lessons as given or can be adapted to fit particular instructional needs. Tasks can also be used as assessments of student understanding of particular Common Core Standards. Or they can be used as a part of professional development to help educators envision the types of instruction and assessment that are faithful to the vision of the new Common Core Standards.

Illustrating and Improving Students' Learning

The primary use of Great Tasks is as a tool for teachers to evaluate the depth of students understanding, to catch misconceptions, and to facilitate classroom discourse designed to extend students' thinking.

Summative Assessment

After the mathematics content has been taught and students have the knowledge needed, these tasks can be used as a final group or individual summative assessment. After the review of the task, teachers will know just what students understand and what concepts need to be revisited. These tasks will also provide information about students' ability to successfully meet the relevant Standards for Mathematical Practice.

Assessment for Learning

When progressing through the material, tasks can be used as assignments along the way to measure students' progress and understanding. The information gathered can be used to guide teacher decisions about future instruction and provide feedback for students to improve their performance.

Professional Development for Teachers

Because the tasks in this collection were developed to illustrate both the philosophy and implementation needed to integrate Great Tasks in the curriculum, they can also serve as models for professional development. For example, presenting these tasks in a workshop can involve discussing the sample tasks and supporting teacher materials before developing similar sets of tasks. The exercise of producing tasks is a nice way to help teachers jump into the essential content of the mathematics that is illustrated by these types of tasks.

Regardless of the audience or purpose, each of these tasks can be used as an individual, partner, small group, or whole group task. Students and educators need the opportunity to participate in each of these types of activities.

- When used for **individual** work, a next step could be to take the time to allow students to work with a partner and evaluate each other's progress. After time spent with a partner, the individual work could be revised and improved.
- The tasks completed with **partners** can be used for discussion about the information as they learn from one another and progress through the task.

- Tasks completed in a **small group** are enhanced when students first read and think about any questions they have before joining the group. Groups can be facilitated by providing roles for each of the individuals. Each group can then share their solution with the entire class. Other groups should be encouraged to ask questions. The learning gained by taking the time to review student thinking helps students develop proficiency in the Standards for Mathematical Practice.

- With any of these methods, a **whole group** discussion of the task can provide cognitive closure as students share how they approached and completed the task. By listening and asking questions, students will learn different approaches to the problems from their peers' work. Planned questions will increase the value of the time spent reviewing the activity. Formative assessments can help improve student achievement by providing time to interact meaningfully and reflect on new information.

Providing Student Work

Another valuable tool to help students learn from a great task is to provide completed work and ask students to evaluate that work. They can look over what has been done to determine if the work is complete; and, in the process, they learn new approaches and begin to enhance their own skills for evaluating their own work.

Rubrics

Although rubrics can be used to evaluate student work and provide a numerical grade, they are just as important for providing specific feedback on critical elements of the task and the student work. Figure 1 provides a generic rubric with descriptions of four levels of accomplishment for each of four characteristics of the work.

Choosing a Rubric

Choose a rubric that focuses on the skills, conceptual understandings, and practices that you want to assess. You might choose to use additional rows if you want to evaluate more than one skill, concept, or practice that a task requires. Although most of the tasks involve several different skills, concepts, and/or practices, not every one that students use in a task needs to be evaluated. The rubric in this publication includes four performance levels; the fourth level reminds us that students at all levels need to be challenged and have opportunities for growth.

Using the Rubric

There are many ways to use a rubric. Often it is helpful when students can see how they will be evaluated before they begin a task. Another approach is to let them complete the task, then review the rubric, and make changes before they submit the work.

A rubric can be used without points attached by highlighting the appropriate sections. If points are a more appropriate use for your classroom, each section (row) can be allocated the appropriate number of points. For example, we could designate three rows for the task's mathematical outcomes, one for the mathematics concepts, two mathematical practices, and finally the last row for communication. The mathematical skills could be set up as 15 points for each row; the conceptual understanding as 20 points; practices could be 15 points each, and 10 points for communication. Then each column could be a percentage of the possible points; Needs improvement 0 to 50%, Approaching Proficiency 51% to 79%, Proficiency 80% to 95%, and Exemplary Distinction 96% to 100%.

However you use the rubric, keep in mind that each one must be tailored to what you want to learn about your students.

Figure 1: Generic Rubric for Assessing Student Work on Great Tasks

	Needs Improvement	Approaches Proficiency	Demonstrates Proficiency	Exemplary Distinction
Mathematics Outcome	Little or no success with the mathematics skill. No workable solution is provided.	Part of the task is correct; however gaps in skill and/or understanding are apparent.	Demonstrates solid mathematical skill presenting a solution, which is correct and complete.	Work demonstrates rigorous mathematical skills and mastery of the concepts, often exceeding expectations.
Conceptual Understanding	Very little understanding of the mathematical concepts involved and/or misunderstood the task.	Some understanding of the relevant concepts is demonstrated.	Demonstrates knowledge of the mathematical concepts involved.	Work shows precise and thorough use of the mathematical concepts critical to successful completion of the task. Special insights or other exceptional qualities are included.
Mathematical Practice	Shows little or no progress toward demonstrating the mathematical practice.	Includes incomplete responses that demonstrate mathematics progress toward the mathematical practice.	Work demonstrates solid mathematical thinking and the ability to successfully use the mathematical practice.	Shows in-depth understanding of essential mathematical practice and eloquence or insight in explaining the practice.
Communication	Writing is confusing or absent.	There is some confusion in the writing and/or charts or diagrams. Mathematics is not clearly explained.	Addresses all processes and components of the task. Explanations are reasonable and clear to the audience.	Writes a comprehensive, compelling, and thoughtful solution. Diagrams are illuminating. Every component of the product is obvious to the audience.

References

Balka, D. S., Miles, R. H., & Hull, T. H. (2011). *Visible Thinking in the K-8 Mathematics Classroom.* Corwin Press.

Bell, C. V., & Pape, S. J. (2012). Scaffolding students' opportunities to learn mathematics through social interactions. *Mathematics Education Research Journal,* 1-23.

Charalambous, C. Y. (2010). Mathematical Knowledge for Teaching and Task Unfolding: An Exploratory Study*. *The Elementary School Journal, 110* (3), 247-278.

Confrey, J., & Maloney, A. (2011, May). Engineering [for] effectiveness in mathematics education: Intervention at the instructional core in an era of common core Standards. In *A paper prepared for the National Academies Board on Science Education and Board on Testing and Assessment for "Highly Successful STEM Schools or Programs for K-12 STEM Education: A Workshop".* Raleigh, NC: The Friday Institute for Educational Innovation, College of Education, North Carolina State University.

Conley, D. (2011). Building on the common core. *Educational Leadership, 68* (6), 16-20.

Common Core State Standards Initiative. (2010). Common core state standards for mathematics. *Retrieved September, 15,* 2012.

Conner, A., Wilson, P. S., & Kim, H. J. (2011). Building on mathematical events in the classroom. *ZDM, 43* (6), 979-992.

Franke, M. L., Webb, N. M., Chan, A. G., Ing, M., Freund, D., & Battey, D. (2009). Teacher questioning to elicit students' mathematical thinking in elementary school classrooms. *Journal of Teacher Education, 60* (4), 380-392.

Gresalifi, M., & Barab, S. (2011). Learning for a reason: Supporting forms of engagement by designing tasks and orchestrating environments. *Theory into Practice, 50* (40, 300-310.

Groth, R. E. (2012). *Teaching Mathematics in Grades 6-12: Developing Research-Based Instructional Practices.* Sage Publications, Incorporated.

National Council of Teachers of Mathematics. (1991). *Professional standards for teaching mathematics.* Reston, VA: Author.

Phillips, V., & Wong, C. (2010). Tying together the common core of standards, instruction, and assessments. *Phi Delta Kappan, 91* (5), 37-42.

Reinhart, S. C. (2000). Never say anything a kid can say. *Mathematics Teaching in the Middle School, 5* (8), 478-483.

Acknowledgements

Texas Instruments Incorporated donated the time and effort of the following individuals to edit and ready these books for publication and add appropriate calculator extensions. We appreciate their dedication.

Elizabeth Bowen, Eric Butterbaugh, Rick Hennig, Gayle Mujica, Jennifer Wilson.

The first writing team worked to get this project started. We want to acknowledge and thank them for their contributions.

Laurie Boswell, Connie Schrock, Richard Seitz, Steve Viktora

We want to thank our reviewer whose comments and input were critical to the shaping of the first tasks.

Steve Leinwand

We want to thank Sue for her feed and help editing some of the later tasks.

Sue Pippin

Cover Design and artwork designed by

Deborah Anker, BesType

Photography by

Connie Schrock

The student work and review of activities were a valuable component in the construction of this document. We apologize for any teachers that are inadvertently not listed here. The authors gratefully acknowledge the contributions of the following individuals.

Avonworth Elementary School, Pittsburg, Pennsylvania: Jennifer Weigand.

Ayers Elementary School, Beverly, Massachusetts: Erin Glencross, Alex Murray, Erin North.

Black Hawk Middle School, Madison, Wisconsin: Kristina Whiting.

Burlingame High School, Burlingame, Kansas: Pam Trimble.

Central Elementary, Helena, Montana: Pam Murnion, Amy Casne-Fetz, Merry Fahrman.

Cherokee Middle School, Madison, Wisconsin: Chris Dyer.

Connersville High School, Connersville, Indiana: Daniel Haffner, Eric Myers.

Connersville Middle School, Connersville, Indiana: Julene Crumley, Donna Litton.

Emporia, Kansas: Helen Williams, Anne Winter.

Emporia Middle School, Emporia, Kansas: Ashley Blome, Morgan Pearson.

Emporia High School, Emporia, Kansas: Casey Collins

Everton Elementary, Connersville, Indiana: Kassindra Young, Jessica Nead, Pam Hermann.

Fayette County School Corporation, Connersville, Indiana: Kathleen Rieke.

Fowler Middle School, Maynard, Massachusetts: Gervase Pfeffer.

Grafton Public Schools, Grafton, MA: Gayle Akillian, Robin Barron, Kelly Carr, Katie Cederberg, Kathy Demartini, Brenda DeVaney, Lauren Eknoian, Deb Fasold, Hilary Kreisberg, Shannon Michalowski, Jane Mason, Christine Papazian.

Hawthorne Elementary School, Helena, Montana: Willy Schauman.

Hawthorne Elementary School, Madison, Wisconsin: Emily Miller.

Leesburg, Virginia: Carolyn Briles.

Lowther North Intermediate School, Emporia, Kansas: Datham Fisher.

Madison Metropolitan School District, Madison, Wisconsin: Laura Godfrey, Katie Wolf.

O'Keeffe Middle School, Madison, Wisconsin: Rob Hetzel.

Olathe High School North, Olathe, Kansas: Cori Samskey.

PIMMS, Cromwell, Connecticut: Mari Muri.

Rock Creek Jr./Sr. High School, Westmoreland, Kansas: Ashley Wege.

Rossiter Elementary, Helena, Montana: Melody Hayes.

Sennett Middle School, Madison, Wisconsin: Becky Walters.

Shaw Elementary School, Milbury Massachusetts: Debra Schroeder, Deborah Lacey, Jane Wojciechowski.

Shawnee Heights High School, Topeka, Kansas: Bradley Nicks.

Spring Harbor Middle School, Madison, Wisconsin: Linda Nelson.

Teachers from Charlotte-Mecklenburg Schools, Charlotte, North Carolina.

Teachers from Kannapolis City Schools, Kannapolis, North Carolina.

Topeka High School, Topeka, Kansas: Renee Chambers.

Village Elementary, Emporia, Kansas: Mayra Vazquez, Traci Meyer.

West Hartford Public Schools, West Hartford, Connecticut: Christine Newman.

White City, Kansas: Joshua Freking.

Robot Arena

6th Grade—Teacher Notes

Overview	
Students will investigate how changes in perimeter affect the area of a figure.	**Prerequisite Understandings** • How to find the perimeter and area of various geometric figures.

Curriculum Content	
CCSSM Content Standards	6.G.1. Find the area of right triangles, other triangles, special quadrilaterals, and polygons by composing into rectangles or decomposing into triangles and other shapes; apply these techniques in the context of solving real-world and mathematical problems. 6.EE.7. Solve real-world and mathematical problems by writing and solving equations of the form $x + p = q$ and $px = q$ for cases in which p, q and x are all nonnegative rational numbers.
CCSSM Mathematical Practices	4. **Model with mathematics**: Students create a model of a rectangle and its dilations to gather data and answer questions. 8. **Look for and express regularity in repeated reasoning**: Students are asked to respond to the question, "Will this always be true?"

Task	
Supplies • Student Worksheet	**Core Activity** Students will work on developing an argument that focuses on a geometric and on an algebraic argument.
Launch Decomposing a figure into rectangles will provide an opportunity to activate students' prior experiences with area and perimeter.	**Extension(s)** Students could be asked to consider geometric and algebraic arguments for three-dimensional figures.

Robot Arena

Launch

Decomposing Figures

Decompose the following figure into two or more rectangles. Find the area and perimeter of the original figure.

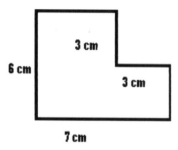

Challenge students to find three different ways to decompose the figure and find the area and perimeter of the original figure. Depending on students' prior experiences, students might be led in a discussion that results in stating the respective formulas for area and perimeter of a rectangle. This examination of multiple methods will provide additional practice and will motivate students to go beyond their first choice.

Area: length x width or $A = lw$

Perimeter: 2 (length) + 2 (width) or $P = 2l + 2w$

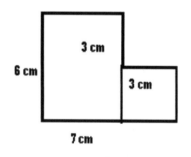
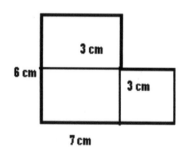
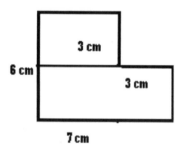

Activity Instructions

Start by allowing the students to work in pairs to come up with at least two rows of data for their table by constructing a rectangle, specifying the dimensions, and then computing the area and perimeter. Students double the lengths of the sides to complete the first row in the table.

Repeat these steps to complete row 2. Once students have an opportunity to come up with 2 entries, ask pairs to share their data until students have several rows completed. (It is not necessary to fill in all of the table rows).

Students can then work in pairs again to explore the remaining parts of the activity.

Extension

When using the extension, students could perform the same activities with a three-dimensional object such as a cube or rectangular prism.

$$\text{Volume} = \text{length} * \text{width} * \text{height} \text{ or } V = l\,w\,h$$

How does doubling the dimensions of the side change the volume? Do you see a pattern? If so, would this pattern hold true for any cube or rectangular solid?

Additional Resources

Appendices A & B contain instructions for **Entering Data** and **Plotting Data** on both the TI-Nspire™ handhelds and the TI-84 Plus graphing calculators.

Robot Arena

Activity

Elan is fascinated with robots. He has two robots that he puts into an arena and experiments with how they move about the space. He wants to design two additional robots and double the space of the arena. His sister Naelly told him that if he doubles the lengths of the sides that the new arena will be twice as large. Elan isn't convinced. Complete the following activity to help Elan decide whether he should follow Naelly's suggestion.

1. Under the table on the next page, construct a rectangle. Record the dimensions in the table on the next page. Find the perimeter and area of the figure and record these in the table as well.

2. Double the lengths of the sides. Record the dimensions, the perimeter, and the area in the table again.

3. Repeat steps 1 and 2 by constructing a rectangle with different dimensions. Record your results in the table.

4. Record examples created by your peers in additional table rows.

5. Think about the effect of doubling the lengths of the sides on the **areas** of the rectangles. Write a description of this relationship.

6. Is this relationship always true? Create an argument using your diagrams and comparing the areas of the pairs of figures. Create an argument using the formula for area.
   Original Area: $A = lw$ Doubled Area: $(2l)(2w)$ or $A = 4lw$.

7. Think about the effect of doubling the lengths of the sides on the **perimeters** of the rectangles. Write a description of this relationship.

8. Is this relationship always true? Create an argument using the diagrams and comparing the perimeters of the pairs of rectangles. Create an argument using the formula for perimeter:

   Original Perimeter: $(P = 2l + 2w)$ or $2(l + w)$ Doubled Perimeter: $2*(2l) + 2*(2w)$ or $4(l + w)$

9. Should Elan follow Naelly's suggestion and double the length of the sides of the arena in order to double the area? Why or why not?

6th Grade — Robot Arena

Dimensions of Rectangles

Dimensions of Rectangle	Dimensions of Doubled Sides	Original Perimeter	Original Area	Perimeter when Lengths Doubled	Area when Lengths Doubled

Rectangles:

Robot Arena

Results from the Classroom

Students have little difficulty finding alternate approaches to the decomposition, but discussion of the results is necessary to demonstrate alternatives. The majority of students in our pilot of the problem had no difficulty finding the area of the original figure as 33 units^2. In general, students experience no problems with finding perimeter; however, several students found the total perimeter of the individual figures from their decomposition and not the perimeter of the original figure (26 units).

For the core part of the activity, students quickly completed the table for Dimension of Rectangles. Interestingly, students' dimensions of choice were overwhelmingly whole numbers. Teachers might want to request students try at least one set of dimensions with fractions.

Dimensions of Rectangles	Dimensions when Sides Doubled	Original Perimeter	Original Area	Perimeter when Lengths Doubled	Area when Lengths Doubled
1, 1	2, 2	4	1	8	4
2, 2	4, 4	8	4	16	16
3, 3	6, 6	12	9	24	36
4, 4	8, 8	16	16	32	64
5, 5	10, 10	20	25	40	100

Students engaged in constructing conjectures about the effect of doubling the lengths of the sides on the area and perimeter of the new figures.

Areas:

> "When the lengths of the sides are doubled, the final area of the rectangle is 4 times the original area."

> "The lengths are doubled and the new area is 4 times what we started with."

Perimeters:

> "When the lengths of the sides are doubled, the final perimeter of the rectangle is 2 times the original perimeter."

> "The new perimeter is twice as much as the original."

The teacher allowed students to work in groups of four for developing justifications of their conjectures. The requirement of a common group response provided excellent opportunities for students to communicate mathematically as they described their mathematical thinking and justified their conjectures.

Two products of these conversations are presented as illustrative of students' abilities to formulate a justification to support their conjectures.

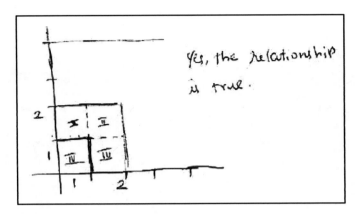

The diagram shows that if the sides are doubled then I get a new rectangle with four similar areas that can fit in the new area. So the new rectangle is 4 x (original area).

Likewise, the students argued mathematically that

$A = l\,w$

Double Area $= 2\,l * 2\,w = 4\,l\,w$

We see that the same discussion supports that doubling gives us 4 times the area (l * w).

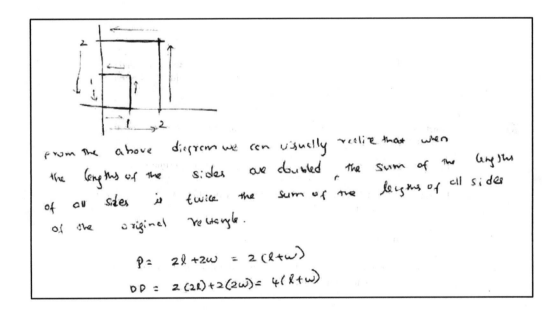

"So, should Elan follow Naelly's suggestion and double the lengths of the sides of the arena in order to double the area? Why or why not?"

Below are a couple of student responses:

"From the evidence, we can suggest that Elan should not follow Naelly's suggestion as doubling the lengths of the sides would increase the area by four times. Instead Elan could double the length of any one side to double the area of the arena."

"Naelly's suggestion would be four times the original area, and Elan only wants it doubled. He needs to measure again to only get twice the area."

The Missing Words

6th Grade—Teacher Notes

Overview	
Students will explore the concepts of variability and distribution of a data set by devising a strategy for determining an unknown quantity.	**Prerequisite Understandings** • Experience with estimation. • Multiple ways to represent data in graphs.

Curriculum Content	
CCSSM Content Standards	6.SP.1. Recognize a statistical question as one that anticipates variability in the data related to the question and accounts for it in the answers. 6.SP.2. Understand that a set of data collected to answer a statistical question has a distribution which can be described by its center, spread, and overall shape. 6.SP.4. Display numerical data in plots on a number line, including dot plots, histograms, and box plots.
CCSSM Mathematical Practices	2. **Reason abstractly and quantitatively:** Students must support their reasoning for obtaining their estimates. 3. **Construct viable arguments and critique the reasoning of others:** Students collect data from their peers and defend their choices.

Task	
Supplies • Student Worksheet	**Core Activity** Students will use a number of strategies for determining the number of missing words from a printed page. They will then write about their methods and conclusions.
Launch Make sure that students have computed measures of central tendency and range. Provide a format for students to report and defend their actions.	**Extension(s)** Locate a copy of *Little House on the Prairie* and share the page referenced in the activity. Students can analyze the strength of their reasoning in estimating the number of missing words.

The Missing Words

Launch

For each condition below, ask students to create a set of five data points that satisfy that condition. Reveal, or read, one condition at a time, and ask several students to share their data and justify their answers before moving on to the next condition. Emphasize that there are many correct answers to each condition.

Condition to Satisfy	Five Data Points	Proof
Mean = 80		
Mean = 80, Range = 10		
Mean = 80, Range = 50		
Mean = 80, Median = 80		
Mean = 80, Median = 100		
Mean = 80, Median = 50		
Mean = 80, Mode = 75		
Mean = 80, Mode = 50, Range = 70		

Activity Instructions

Assure students that this is an actual page from the book, *Little House on the Prairie*. In providing lines to write a letter, students find it much easier to actually write words versus simply showing mathematics. Student thinking is more transparent.

Additional Resources

Appendices A & B contain instructions for **Entering Data** and **One-Variable Statistics** operations on both the TI-Nspire™ handhelds and the TI-84 Plus graphing calculators.

The Missing Words

Activity

Hitoshi began reading *Little House on the Prairie* after he bought the book at the Reading Times Bookstore. When he got to page 332, he noticed that there was something wrong. He went to the bookstore to complain, but the store manager would not give Hitoshi a refund or a new book. However, the manager finally agreed to give him a new book if Hitoshi could provide both:

- a **reasonable estimate** of how many words were missing, and
- a **clear explanation of how he had figured it out.**

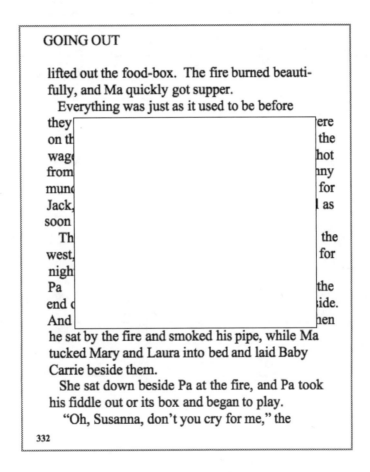

Part 1

Your Job

Write the letter below that tells Hitoshi what he can say to the manager. Be sure to include:

- about how many words do you think you are missing, and
- how you figured out how many words were missing.

Don't worry too much about spelling. Just do your best. Hitoshi is your age and will understand that this is just your "first draft."

Dear Hitoshi,

I think there are about _____ words in the missing section of the page. Here's how I figured it out.

_____

_____

_____

_____

_____

_____

_____

Good luck,

_____

Part 2

In order to answer the following questions, you will need to have the estimates of the missing numbers of words from all of your classmates.

1. Collect the data from your classmates and list them in the space below.

2. Find the mean of the estimates of the missing words.

3. What is the range of estimates? Why do you think this number is as large as it is?

4. Create a graph representing the information. Explain why you made the graph you selected.

5. Decide how to identify the mean on your graph. Does your graph tell you any additional information about the set of data (classmates' answers)? Explain.

The Missing Words

Results from the Classroom

Part 1

Rebecca

Rebecca took a very basic approach to her estimate and was able to easily explain it.

> Dear Hitoshi,
>
> I think there are about __100__ words in the missing section of the page. Here's how I figured it out.
>
> I took all of the words that were there and I estimated how many words would be under there. So I imagined there were 10 words in each sentence and there where 10 sentences so I multiplyed them 10x10 and I got 100 words missing.

Sari

Sari explained where she got her estimate of 10 for the number of words and then estimated more lines missing than Rebecca did.

> Dear Hitoshi,
>
> I think there are about __130__ words in the missing section of the page. Here's how I figured it out.
>
> Because the sentence above it was 9 + right below it was eleven, + the average is 10 so I used the number of sentences that is missing + multiplyed it + got 130

Chase

Chase has the largest estimate for the missing words by a larger estimate for the length of the line and the number of missing lines.

Dear Hitoshi,

I think there are about __169__ words in the missing section of the page. Here's how I figured it out.

I counted the amount of words there was and i just times 13 by 13 and got 169 words.

☺

Adriana

This is a different approach to the estimate and provides one of the smallest estimates. It is good when students can share a different strategy for producing the estimate.

Dear Hitoshi,

I think there are about __75__ words in the missing section of the page. Here's how I figured it out.

I counted how many words were in the sections you can see. I added them together & got 75. The two numbers I added together were 52 & 23. I did it this way because, I think the missing section is about the same size as the other to sections put together.

Part 2

The second part of the task involved collecting the data and formulating an interpretation of the data. It is more important as it related to the practices; unfortunately, the student answers are missing some of the depth that we would like to see.

3. What is the range of estimates? Why do you think this number is as large as it is?

169
75
94 -Range

Because their was so many words.

3. What is the range of estimates? Why do you think this number is as large as it is?

That what they feel it is.

169
75
094

3. What is the range of estimates? Why do you think this number is as large as it is?

mean, 107÷9 =

@ 94
I think it is this large because we did the math.

169
-75
94

3. What is the range of estimates? Why do you think this number is as large as it is?

94) 169
    - 75
      094

Because the space is bigger than the other spaces, so the number should be bigger than the other spaces

4. Create a graph representing the information. Explain why you made the graph you selected.

6th Grade   36   The Missing Words

When asked to make a graph and explain why they selected the one they did, it was again clear that more work is needed to help students at this level of development to justify their selection. Many different types of graphs were created, and the students clearly demonstrated the ability to create the graphs.

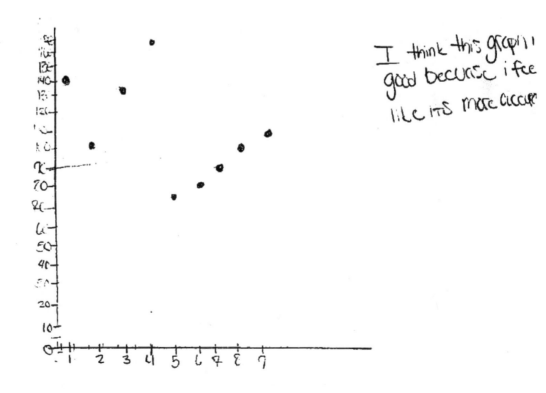

A line plot is a good graph because it shows all the numbers & it's easy to see the numbers placed on the line.

4. Create a graph representing the information. Explain why you made the graph you selected.

I think this graph is good because i fee like its more accua[te]

4. Create a graph representing the information. Explain why you made the graph you selected.

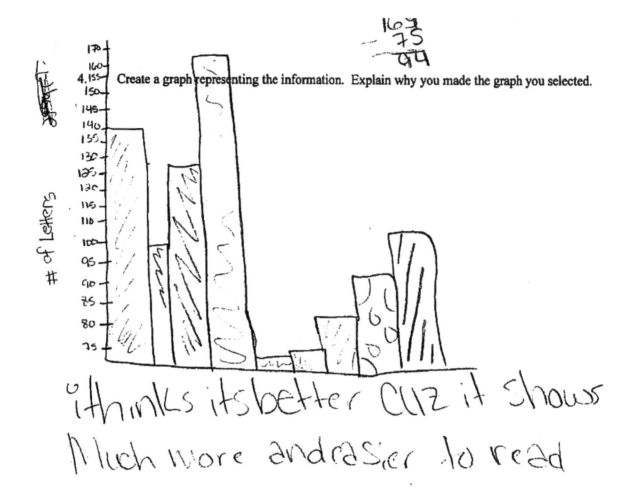

i thinks its better cuz it shows Much more and easier to read

How Many Fit?

6th Grade—Teacher Notes

Overview	
Students will use fraction strips to develop a conceptual understanding of division of fractions.	**Prerequisite Understandings** • Work with fractions on a number line. • Multiplication of fractions using area models or arrays.

Curriculum Content	
CCSSM Content Standards	6.NS.1. Interpret and compute quotients of fractions and solve word problems involving division of fractions by fractions, e.g., by using visual fraction models and equations to represent the problem.
CCSSM Mathematical Practices	2. **Reason abstractly and quantitatively**: Students make sense of fraction operations with a concrete model (fraction strips). 7. **Look for and make use of structure**: Students observe patterns in their work and the relationship to the problems posed. 8. **Look for and express regularity in repeated reasoning**: Students are asked to generalize from the patterns in the problems and their work.

Task	
Supplies • Fraction strips (Commercial or homemade)	**Core Activity** Students apply a real-world situation to an abstract concept to better understand the patterns of fractions.
Launch Explore the use of fraction strips for representing fractions, equating them, and multiplying and dividing unit fractions.	**Extension(s)** Develop different procedures for the division of fractions based on their multiplication experience in the activity.

How Many Fit?

Launch

Teaching Notes:

Many commercially developed fraction strips label each strip in terms of its size compared to the whole. This enables students to read the size rather than having to reason about the size of each strip. Cover the labels with masking tape as we want students to continue to evaluate the size of the strips in relation to each other and keeping the whole as the reference.

Every student should have a set of fraction strips to use. Ensure that students understand the representation of all of the pieces by asking students to hold up the piece that is the size of 1/4, then 1/8, etc. Ask students to show you equivalent fractions, perhaps working with a partner to represent as many equivalent fractions as they can using fraction strips.

Continue the launch of this lesson by asking students to complete the Launch Worksheet that focuses on multiplication and then division of unit fractions.

Part A

Examples of fraction strips:

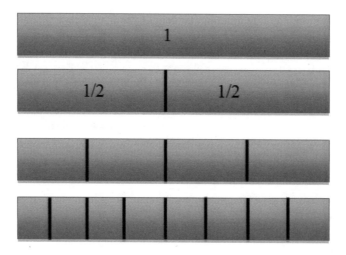

1. What is 1/2 of 1/4?
2. What is 1/4 of 1/2?
3. Compare your answers to #1 and #2. What do you notice?
   - What property is being used?
4. Find 1/8 of 1/2.
5. Find 1/2 of 1/8.
6. Compare your answers to #3 and #4. What do you notice?
   - What property is being used?
7. Find 1/2 of 3/4.

8. Find 1/2 of 5/8.
   - How are problems #7 and #8 different from those in #1, #2, #4, and #5?
9. Compare your answers and share your reasoning as to whether your answers make sense. Be ready to explain your thinking.
10. How would you solve these problems if you did not have fraction strips?
    - Write down the steps that you would take to solve these problems.
11. Select one of the problems from above. Write a story that would be solved by using the problem that you selected.

Part B

Work independently to solve these problems using your fraction strips:

1. How many pieces the size of one-fourth fit in 1/2?
2. How many pieces the size of one-eighth fit in 1/2?
3. How many pieces the size of one-sixteenth fit in 1/2?
4. Look back on your three answers. What pattern do you see?
   - Explain why you think this pattern is happening.
5. How many pieces the size of 1/8 fit in 1/4?
6. How many pieces the size of 1/8 fit in 1/8?
7. How many pieces the size of 1/8 fit in 1/16?
8. Look back on answers to questions #5 - #7. What pattern do you see?
   - Explain why you think this pattern is happening.

Work with your partner:

9. Compare your answers to Part B. Be ready to justify your reasoning.
10. Select one of the problems in # 1-3 or #5-7. What operation is indicated in these examples?
11. Write an equation using the values in your selected problem.
12. Write a story problem that would use your equation to find the solution.

Discuss

After completing these two pages, hold a group discussion about their conclusions.

In Part A, listen for students to discuss the answers to # 1 and #2 and #4 and #5 being the same because of the commutative property. In #7 and #8, the problems involve fractions that are not unit fractions. Highlight for students that they know that 1/2 of 1/4 is 1/8 and that 3/4 is the same as 3 times 1/4, so the answer to 1/2 of 3/4 must be three times larger than 1/2 x 1/4.

Students might be puzzled by the fact that they are multiplying, yet the answers are smaller than the factors. If this misconception is raised, use an area model to help students visualize what is happening.

In Part B of the Launch, students are working with division to explore how many pieces fit. In #1-3, students are asked to reflect on the pattern that the answers are doubling each time. Focus students' attention on the divisor. This divisor is being cut in half each time. When the divisor is cut in half, the solution doubles. Ask students to consider the size of each of the answers.

As the problem uses fractions, why are the answers apparently whole numbers? This is the objective of the Core Activity.

The discussions following each section of the activity will engage students in finding patterns that will bridge toward the standard algorithm.

Extension

Students can begin to think about different procedures that build on this foundational understanding of division of fractions. Present them with the following:

We discovered that we could multiply fractions by multiplying the numerators together and then multiplying the denominators. For example: 3/4 x 1/2 = 3/8.

Do you suppose that we can use a similar procedure for division: 3/4 ÷ 1/2 = ?

Can we divide 3 by 1 and then 4 by 2? Does that lead to the correct answer? Use your fraction strips to verify whether or not this is true.

How Many Fit?

Activity

From the Launch:

> 5) How many pieces the size of 1/8 fit in 1/4? _____
>
> 6) How many pieces the size of 1/8 fit in 1/8? _____
>
> 7) How many pieces the size of 1/8 fit in 1/16? _____

Two students are discussing the problems that they just completed in Part 2 of the introduction. They are both struggling to understand whether or not their answers make sense.

Pedro says, "Let's try to come up with a situation where we'd have to use these fractions."

Netch responds, "You think that we might be able to make some sense of this if we know the situation? Okay, let's try."

Pedro suggests that they think about pizzas. "Suppose we have one-half of one pizza. We want to share this pizza among 4 of us. How much of the pizza will each person get?"

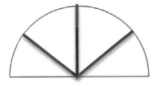

1. Help Pedro and Netch solve this problem. Be certain to show your work and be ready to explain your thinking.

Netch and Pedro are continuing to think about division of fractions. So, let's now look at some patterns.

2. In our last problem, we had 1/2 pizza. We wanted to share with 4 people. So we cut the half of the pizza into four equal parts. What size is each person's share?

   ½ ÷ 4 = _____

3. Fill in the blanks below. Use a diagram or fraction strips to help you.

   ½ ÷ 2 = _____

   ½ ÷ ½ = _____

   ½ ÷ ¼ = _____

Netch says, "Look. There is another pattern!" In each of your problems, the divisor or the number of shares we are looking for is being cut in half. Look at the quotients!"

4. What is the pattern that Netch sees?

5. Explain why you think that this pattern occurs:

How Many Fit?

Results from the Classroom

Students in this sixth grade class had used fraction strips earlier in the year. Several students decided that they didn't want to use the strips until they began working on the core task. They then found the strips helpful as they looked for patterns. The class period was shorter than usual, so a majority of students did not complete the task in the 40 minute period.

Launch

Students readily completed this section; the only difficulty arose when they were asked about the property being demonstrated. Most students skipped that question. Students had more difficulty in Part B.

For questions #1 – 3, Jacob expressed his answer as a fraction rather than the number of pieces. The numerator he used does equal the number of pieces; he recognized that he had written equivalent fractions as he reflected on the pattern.

> Work independently to solve these problems using your fraction strips:
>
> 1) How many pieces the size of one-fourth fit in ½? __2/4__
>
> 2) How many pieces the size of one-eighth fit in ½? __8/8__ (?)
>
> 3) How many pieces the size of one-sixteenth fit in ½? __8/16__
>
> 4) Look back on your three answers. What pattern do you see? __it was equal to each other__

Melanie connected these same questions with repeated subtraction. She wrote for an explanation of the pattern, "It goes by 2's, and it goes in order when you subtract by 2's."

In responding to "How many pieces the size of 1/8 fit in 1/16," several students wrote "0." They stated that the 1/8 was just "too big."

Mallique thought additively. She also told her teacher that she didn't need to use the strips, but her teacher encouraged her to check her work using the strips during the next class.

> Work independently to solve these problems using your fraction strips:
>
> 1) How many pieces the size of one-fourth fit in ½? __2__
>
> 2) How many pieces the size of one-eighth fit in ½? __4__
>
> 3) How many pieces the size of one-sixteenth fit in ½? __6__
>
> 4) Look back on your three answers. What pattern do you see? __for every forth you add you go up by 2__
>
> Explain why you think this pattern is happening? __if you add a forth you go up by two__

Activity

Many students answered the first question on the core task correctly. The successful students drew diagrams to record their thinking.

AJ performed the division of fractions using the standard algorithm. It is interesting to note that he used the algorithm on $\frac{1}{2} \div \frac{1}{2}$. His teacher asked him to reflect on this problem again. AJ responded, "Oops. Sometimes I race ahead without thinking."

Help Pedro and Netch solve this problem. Be certain to show your work and be ready to explain your thinking.

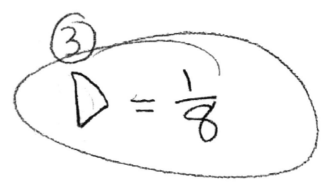

Sizing Up the Garage

7th Grade—Teacher Notes

Overview

Using the Hot Wheels® scale factor of 1:64, students will design a garage by calculating with scale factors and drawing and constructing scale models.	**Prerequisite Understandings** • Understand ratio concepts. • Ability to use ratio reasoning to solve proportions.

Curriculum Content

CCSSM Content Standards	7.G.1. Solve problems involving scale drawings of geometric figures, including computing actual lengths and areas from a scale drawing and reproducing a scale drawing at a different scale.
CCSSM Mathematical Practices	4. **Model with mathematics**: Students calculate the dimensions of a full-size model from a scale model car. 5. **Use appropriate tools strategically**: Students choose their tool for measuring and their tool for scaling up to full-size. 6. **Attend to precision**: Students will have to manage the precision of measuring with a ruler and converting a decimal to the nearest tenth.

Task

Supplies	Core Activity
• Hot Wheels® or similar model/toy cars • Standard ruler • Paper	Students measure model cars and convert scale measurements into full-size measurements. Then have students create a floor plan of a two-car garage continuing to use the 1:64 scale factor.
Launch	**Extension**
Use either the given line segments or classroom objects to practice measuring length and scaling figures. An additional practice component of converting their resulting measurements into decimal equivalents will be helpful as well.	Students could design a 3-dimensional garage model based on their floor plan and place their car in it to verify whether or not it will fit in the garage.

Sizing Up the Garage

Launch

Measuring Length

Measure the following line segments using a standard ruler. Convert the fractions into decimals, rounding to the nearest tenth.

Scaling Figures

Before beginning Scaling Factors, review the meaning of similarity with your students. Students should use the given scale factor (or another that you choose) to draw similar shapes.

Draw a similar figure for each of the shapes given using a scale factor of 5:4.

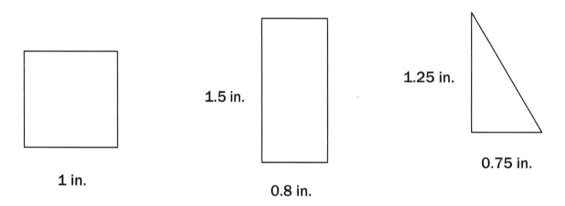

Introduce Activity

Begin by asking students to think about building a garage—how big should it be? How would they find out? If two cars simply fit in the garage, is that good enough? What else should they consider? Can it be too big? What if costs are limited? This kind of a discussion will help them better understand criteria they should follow later when evaluating the size of the garage in comparison to their car and while designing their own garage. After they have finished their garage designs, ask students to explain how they know their proposed garage will fit two of their cars.

While measuring the car, accuracy will be very important. Students could verify a partner's measurements for improved accuracy. It is recommended that students convert their measurements into decimal form.

Sizing Up the Garage

Activity

Determine if two full-scale versions of your car would appropriately fit into an 18' x 20' garage.

1. Measure your car. Round your answers to the nearest tenth of an inch.

   a) Length _____ inches

   b) Width _____ inches

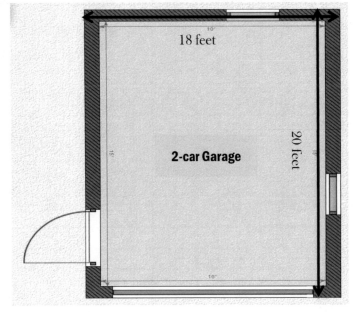

2. Using the Hot Wheels® standard scale factor of 1:64, what are the actual measurements of your car? Be sure to show all of your work below to explain how you found your answer. Round your answers to the nearest tenth of a foot.

   - Is rounding to nearest tenth consistent with using customary measure?

   a) Length _____ feet   b) Width _____ feet

   c) Will the full size version of your car fit appropriately in the 18' X 20' garage? Briefly explain how you know.

3. Draw the floor plan of a two-car garage in which two of your cars would fit appropriately.

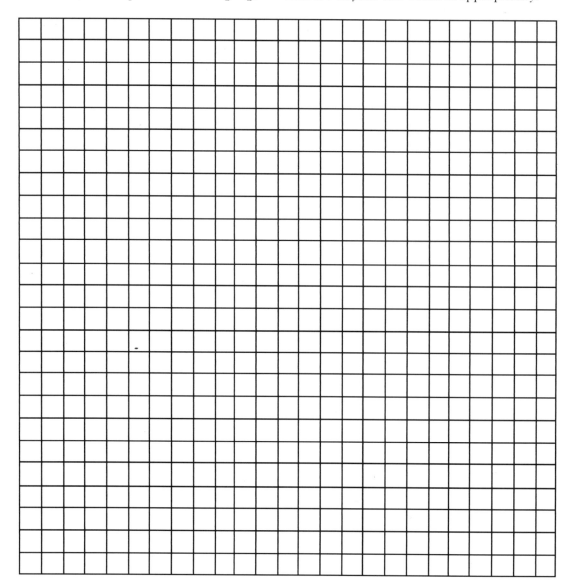

4. Write a note to the general contractor who will build the garage from your floor plan explaining how you know two of your cars will fit.

Sizing Up the Garage

Results from the Classroom

The measurements students provided of their cars varied greatly. Many students avoided any decimals and simply recorded the measurements as whole numbers. Teachers might need to provide support as students measure the cars. One teacher suggested completing an initial whole class activity which engages students in measuring a few items and recording the measurements in the nearest tenth of an inch.

While the car measurements will vary depending on the types used, the following illustrates an acceptable response.

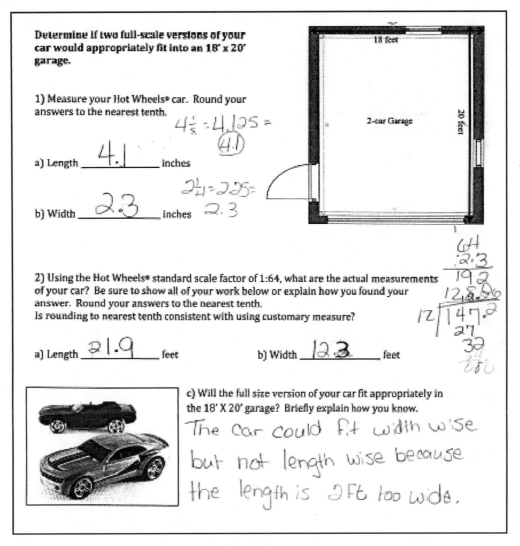

The majority of students easily concluded that the garage was not sufficient because the length of the actual cars, using a scale factor of 1:64 would result in a structure with an insufficient length. Errors in reaching this conclusion were generally traced back to difficulty with converting measurements from inches (for the car model) to feet.

In general, students did not do as well with part 3 of the activity. Their floor plans were often not drawn with any scale in mind. The grid should prompt students to use the squares as a scale for their floor plan.

For example, one student indicated that one square (□) = 1 ft. and drew a diagram showing a garage with dimensions 25 ft. (25 squares) by 16 ft. (16 squares). For many diagrams, however, there was no evident scale.

Part 3 provided rich discussion around how much space would one need for two cars. For example, students argued that the 25 ft. by 16 ft. garage would not hold two cars measuring 21.9 ft. by 12.3 ft. Students realized that they needed to double the dimensions (i.e., 43.8 ft by 24.6 ft), but their responses for the dimensions of the two-car garage often used these figures which left no room between or around the cars. Students were quick to realize this error of thinking, but their responses too often demonstrated a lack of precision and not considering real-life factors in making decisions about the dimensions of the garage.

This common problem is represented in the following student response to part 4 (writing a note to the general contractor explaining how one knows the two cars will fit).

> Dear General Contractor,
> If one of my cars is 13.7 x 6.6 then I know that two of my cars are 27.4 x 13.2. So, I need my garage to be at least those demintions.

Hot Wheels® is a registered trademark of Mattel, Inc.

Most Square?

7th Grade—Teacher Notes

Overview	
Students will explore ratios by comparing the lengths of the sides of properties to determine which is most square.	**Prerequisite Understandings** • Characteristics of a square, particularly that the lengths are equal.

Curriculum Content	
CCSSM Content Standards	7.RP.2. Recognize and represent proportional relationships between quantities. 7.G.6. Solve real-world and mathematical problems involving area, volume and surface area of two- and three-dimensional objects composed of triangles, quadrilaterals, polygons, cubes, and right prisms.
CCSSM Mathematical Practices	2. **Reason abstractly and quantitatively**: Students use relationships between quantities to reason about ratios. 3. **Construct viable arguments and critique the reasoning of others**: Students justify their mathematical thinking and compare their approaches to an approach presented in the problem.

Task	
Supplies • Rulers • Graph paper	**Core Activity** Students explore a problem considering the dimensions of three rectangular areas and deciding which is most square. The focus of the activity is for students to realize that the sides of a square form a ratio that can be simplified to 1 unit.
Launch The launch activity will review basic concepts about ratios and proportions and engage students in using proportions to solve a real-world problem. In addition, review the characteristics of a square.	**Extension(s)** Have students create similar problems that use information to form ratios that would then be compared to the ratio of 1:1. Students might also be asked to use percents in developing arguments for their responses.

Most Square?

Launch

A **ratio** compares two numbers or quantities. Ask students to generate examples of ratios. Collect examples or add ones until the three types of ratios are presented: part to part, part to total, and total to part. Also reinforce different ways that ratios might be written.

Part to Part	●●●●●●●	4 to 3 4 : 3 $\frac{4}{3}$
Part to Total	●●●●●●●	4 to 7 4 : 7 $\frac{4}{7}$
Total to Part	●●●●●●●	7 to 4 7 : 4 $\frac{7}{4}$

This activity will also require students to think about proportion in determining which rectangular lot dimensions are 'most square'.

For example, if you were driving at 60 miles an hour, what distances could you expect to travel over various time periods?

Hours	1	1.5	2	2.5	3
Miles	60	90	120	150	180

This type of thinking supports proportional reasoning as the table reinforces the idea that proportions involve two things that both change at the same rate.

Students should be engaged in discussing proportions and which proportion would form a square. For example, a square with sides 16 cm would give a ratio of 16:16 or 1:1 for the sides. This is the basis of students' proportional thinking as they work through the activity. Percentages and comparing fractions are also emphasized.

If student pairs have difficulty making the initial observation about the ratio of the sides of a square, the teacher might consider providing guidance to individual pairs or engage the class in a brief discussion so that solving the task is facilitated. The lot dimension closest to a ratio of 1:1 is the 'most square'.

Most Square?

Activity

A local business has four empty lots available for commercial development:

- 195 feet by 250 feet
- 278 feet by 315 feet
- 80 feet by 115 feet
- 355 feet by 406 feet

A potential client asked which lot was 'most square'.

How would you answer the question? Explain your thinking.

Prepare a justification to present to the class. You might want to visit with a partner to discuss how to present your ideas. You also might want to describe all the properties of a square and describe how closely these 'square imposters' come to meeting those criteria.

Most Square?

Results from the Classroom

Launch

Students will likely have little difficulty with the launch activity though teachers will need to continue prompting students with examples of ratios until examples of all three types have been offered.

Students offered the following examples from their classroom:

- 6 goldfish to 11 guppies in the aquarium is an example of Part to Part, 6 : 11
- 6 out of 22 students are wearing red is an example of Part to Total, 6 : 22
- Of the 12 boys in class, 9 have a brother or sister is an example of Total to Part, 12 : 9.

Core

The core activity provided insights into students' proportional thinking. A number of students used an approach that looked for the smallest difference between the lengths of the two sides – recognizing that the difference between the sides of a square is zero. Notice in the student example below how the differences in the lengths of the sides appear in the interior of the rectangles.

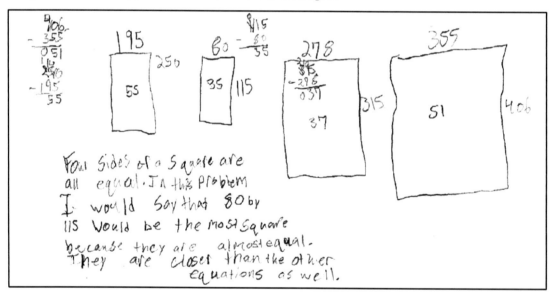

In the discussion at the end of the activity, the teacher presented students with a rectangle that measured 5 by 8 (difference of 3). She doubled the sides to 10 by 16 (difference of 6) and asked if the smaller rectangle was 'more square' than the larger one. She also reinforced that these two rectangles were similar figures. This line of questioning was successful in challenging students to reconsider the 'differences in side lengths' as a viable approach.

Some students also formed ratios of the two sides: 195/250, 278/315, 355/406, and 80/115, concluding that the rectangle with dimensions 278/315 is closer to a square because the ratio is closer to one. One pair of students converted the ratios to decimals arguing that a square's sides would be 1.0, so the rectangle closest to 1.0 would be 'most square'. After the class discussion, students were able to revise their justifications. The teacher encouraged them to use ratios and proportions in revising their justifications.

Odd or Even

7th Grade—Teacher Notes

Overview

Students are asked to settle an argument between two people about the probability of either one winning a contest.	**Prerequisite Understandings** • Ability to calculate basic probabilities. • Definitions of even and odd numbers.

Curriculum Content

CCSSM Content Standards	7.SP.6. Approximate the probability of a chance event by collecting data on the chance process that produces it and observing its long-run relative frequency, and predict the approximate relative frequency given the probability.
CCSSM Mathematical Practices	3. **Construct viable arguments and critique the reasoning of others**: Students explain who is correct and explain the results based on the mathematics. 4. **Model with mathematics**: Students use probability to simulate the situation.

Task

Supplies • Spinners (manual and/or electronic)	**Core Activity** In pairs, students will explore the experimental probabilities and calculate the theoretical probabilities of even and odd sums of random numbers.
Launch Briefly review the definitions of even numbers and odd numbers and probability. Specifically discuss that 0 is an even number. Consider patterns from the number line and the definition of even.	**Extension(s)** Ask students to work in groups and create a "fair" game where each player would have an equal chance of winning. Make sure they can explain why it is fair.

Odd or Even

Launch

Spinner A will be the three-part spinner and Spinner B will be the four-part spinner. Spin both spinners, and record the results in the chart below. Create a fraction in the next column, writing the result as a rational number $\frac{A}{B}$. Then, in the remaining columns, simplify your fraction, if possible, and find the decimal and percent equivalents. Repeat this 5 times (or for 5 "trials").

Trial #	Spinner A	Spinner B	Fraction $\frac{A}{B}$	Simplified Fraction	Decimal	Percent
1						
2						
3						
4						
5						

In the following tables, list all possible outcomes of the rational number $\frac{A}{B}$. Then simplify and find their decimal and percent equivalents. (Hint: There are 12 possible outcomes.) (Hint #2: You already have some of these done in the first table!)

$\frac{A}{B}$	Simplified Fraction	Decimal	Percent

$\frac{A}{B}$	Simplified Fraction	Decimal	Percent

Organize your distinct **outcomes** $\frac{A}{B}$ (as simplified fractions) from least to greatest in the first column of the table shown to the right.

Find the **frequency** and the **probability** of each outcome and record each in the remaining columns.

Outcomes	Frequency	Probability

If we were to play a game where Player 1 gets a point when the outcome is greater than or equal to 1 and Player 2 gets a point for an outcome less than 1, which player would you want to be and why?

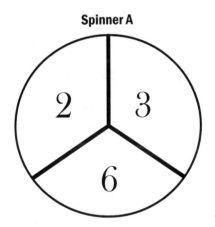

Spinner A

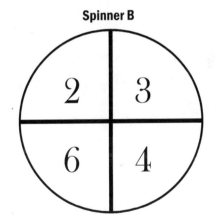

Spinner B

Additional Resources

Appendices A & B contain instructions for **Generating Random Integers** on both the TI-Nspire™ handhelds and the TI-84 Plus graphing calculators.

7th Grade — Odd or Even

Odd or Even

Activity

Leo and Tarra are playing a spinner game with the following rules:

When it is a player's turn, the player spins both spinners. They then find the sum of the two numbers. If the sum is EVEN, player 1 wins (Leo). If the sum is ODD, player 2 wins (Tarra).

Leo takes a test spin first. Here is what he spins:

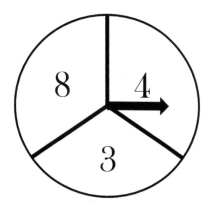 

The sum from the first spin is EVEN because 4 + 0 is even. Leo wins. Leo says, "I like this game. I have a better chance to win it than you do."

Tarra says, "No, I have a better chance to win it than you do."

1. Use mathematics to decide which player is correct.

2. Write a note to the players explaining how you know who has the better chance of winning.

Odd or Even

Results from the Classroom

Olivia

Olivia understood and was careful not hurt Leo's feelings by saying he was partially right. She was able to answer the probabilities with precision.

> Dear Leo- I belive you are wrong and right. If you are talking about theoretical probability you are correct. Because you have 5/9 possibilities as opposed to Tara who has a 4/9 chance.

Latisha

Latisha has a clear and complete understanding of problem. Her explanation distinguishes between theoretical and experimental probabilities. Her argument is correct.

> Leo,
>
> You are correct, theoretically speaking. The chances that an even sum will be spun is 5/9. Tarra could win. however, you could do many test trials to find the experimental probability of even or odd outcomes, which might be different than the theoretical probability. Yes the chances say you will win, but that can change.
>
> Sincerely,

> Dear Leo + Tarra,
> Whoever has the even numbers has a $\frac{5}{9}$ chance of winning with theor[etical] probability. With experimental though, I d[on't] know who would win, I would [want] to look at the data for previous spins. Leo is <u>probably</u> going to wi[n], not for sure, just probably! So good luck to you guys in your game, and, go Leo!

Juan

Juan talks about looking at the data and uses the term probably to describe Leo's chance of winning. He is more specific by mentioning that the even-number player will win. All of the students have been motivated by the situation and have written specifically to the learners.

Isabella

Isabella is specific with her information and lists the possibilities for each person to win. She extends the information without being asked to explain how she thinks that the players could make it a fair game. She demonstrates both of the highlighted mathematical practices. She has modeled the situation with mathematics and was able to present a clear mathematical argument.

> Dear Leo and Tarra,
> I'm sorry but Leo has a better theoretical possibility to win the game. While Tarra only has 4 possibilities, Leo has five. You would have to make it fair though, you could take turns on whoever gets to be player one.
>
> P.S. Tarra, here are your possibilities: 3+8
> 8+1
> 4+1
> 3+4
>
> Leo, here are your possibilities: 4+8
> 8+8
> 8+1
> 8+4
> 4+4
>
> See, there are more for Player 1!
> (Sorry player 2)

7th Grade — Odd or Even

How Do You Visualize It?

8th Grade—Teacher Notes

Overview	
Students explore functions, describe patterns, and discover ways that functional representations can describe a pattern.	**Prerequisite Understandings** • Using variables to describe quantities and construct simple equations.

Curriculum Content	
CCSSM Content Standards	8.F.4. Construct a function to model a linear relationship between two quantities. Determine the rate of change and initial value of the function from a description of a relationship or from two (x, y) values, including reading these from a table or from a graph. Interpret the rate of change and initial value of a linear function in terms of the situation it models, and in terms of its graph or a table of values. 8.F.5. Describe qualitatively the functional relationship between two quantities by analyzing a graph (e.g., where the function is increasing or decreasing, linear or nonlinear). Sketch a graph that exhibits the qualitative features of a function that has been described verbally.
CCSSM Mathematical Practices	2. **Reason abstractly and quantitatively:** Students describe how multiple functions can be used to describe a pattern. 4. **Model with mathematics:** Students apply their knowledge of functions to solve real-world problems.

Task	
Supplies • Graph paper and chart paper • Markers/pens • Graphing calculators	**Core Activity** Students explore which functions correctly describe a given situation to better understand relationships between functions and the situations they represent.
Launch Use the introduction to review the definition of a variable and a function.	**Extension** Have small groups develop their own pattern for a pool border and follow the activity.

How Do You Visualize It?

Launch

Variables and Expressions

As a launch activity, ask students to write down the meaning of *variable*. Have students share their meanings.

Two important ideas should be reinforced:

1. Variables are letters or symbols which represent an unknown value.
2. Variables might change (or vary) based on the constraints of the information presented.

Students should understand two common roles of variables in expressions. Put the following two expressions on the board and ask students in small pairs to discuss how variables are used.

$$2x + 1 = 7$$

$$1 = x\left(\frac{1}{x}\right)$$

In the first expression, the variable might be best thought of as a placeholder for a specific number. In this expression, x has one value (x=3). In the second expression, the variable x represents a pattern or generalization. The product of a number and its reciprocal is equal to 1.

Next, present a word problem to ground their thinking about variables and expressions.

At Sweet Sensations, the price of a cup of frozen yogurt depends on the number of toppings selected. Which of the following could represent this relationship? Justify your thinking.

A. $y = .8x^2$

B. $y = .5x - .3x^2$

C. $y = .8x + 4$

D. $y = \frac{1}{2}x + 4$

Functions

Present the definition of a function. A function describes the relationship between two quantities where the value of one variable depends on the value of the other.

Graphing a function that represents a concrete situation helps students explore the power of functions to represent situations which vary. Students use their graphs to find the number of tiles for various values of *n*. The input/output nature of functions reinforces the role of variables and the nature of functions in representing situations.

How Do You Visualize It?

Activity

Naelly and Elan are excited about updating their outdoor pool which includes square tiles around the perimeter of the pool as illustrated in the diagram below. In describing the project, Naelly commented "I visualize the tiles around the pool as $4n + 4$ squares." Elan added, "No, I think it is more like $2(n+2) + 2n$."

1. Which is correct? Why?

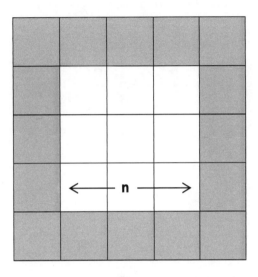

2. Work with a partner to find another function to describe how one might visualize the pattern of tiles around the pool.

3. The pattern could work for pools of various dimensions. Graph one of the functions using your graphing calculator. Use the graph to discuss how many tiles would be needed for the border for pools of various dimensions. List the dimensions you use.

How Do You Visualize It?

Results from the Classroom

Defending an individual's mathematical thinking appeals to students and engages them actively in thinking about problem solving and justifying mathematical ideas. Working in groups facilitated discussions of the approaches as students talked about how they might 'figure out who's right". The teacher facilitated the discussion by asking groups to make sure they considered both Elan's and Naelly's approaches.

The response below is interesting in that it includes two ideas for seeing which approach was correct. First, the students try a value for n, verifying that both of the approaches result in 16 tile when $n = 3$.

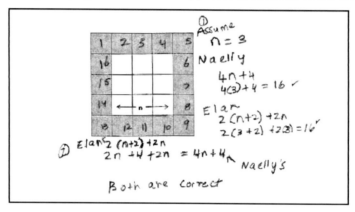

In part 2 of the response, note that Elan's $2(n + 2) + 2n$ is expanded and simplified to show that it is equivalent to Naelly's approach. Both of these methods were part of the whole-class debrief.

The teacher engaged the students in talking about how they could visualize $4n + 4$ as a pool that is 3 tiles by 3 tiles ($n = 3$), which would have 4×3 for the border at the pool and 4 tiles for the corners. Students visualized $2(n + 2) + 2n$ by taking two sides with $n + 2$ tiles (that is, n for the tiles directly at the pool and 2 tiles for the corners), leaving n tiles on the other two sides. Discussing how the two different, yet equivalent, functions can describe a relationship was fruitful for the entire class.

One possible function offered for visualizing the pattern of tiles around the pool was $4(n + 1)$. It is important for this part that students describe how the function represents their visualization of the pattern. The students offering this function circled the following groups of tiles (1,2,3,4), (5,6,7,8), (9,10,11,12), and (13,14,15,16). This showed how each group was n (3 tiles directly at the pool) + 1 additional tile.

Finally, students graphed one of the functions and discussed how many tiles would be needed for borders with various dimensions. In the example below, students also concluded that $4n + 4$ represents the relationship, which they identify as linear.

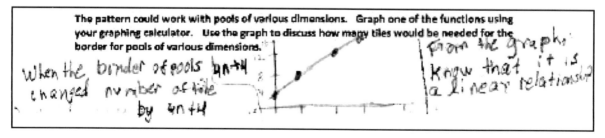

The Shrinking Square

8th Grade—Teacher Notes

Overview

	Prerequisite Understandings
This task uses a geometric pattern and the Pythagorean Theorem to find and to generalize patterns involving radicals.	• Experience working with radicals and rational and irrational numbers. • Experience finding area and perimeter of squares, and using the Pythagorean Theorem.

Curriculum Content

CCSSM Content Standards	8.EE.2. Use square root and cube root symbols to represent solutions to equations of the form $x^2 = p$ and $x^3 = p$, where p is a positive rational number. Evaluate square roots of small perfect squares and cube roots of small perfect cubes. Know that $\sqrt{2}$ is irrational. 8.G.7. Apply the Pythagorean Theorem to determine unknown side lengths in right triangles in real-world and mathematical problems in two and three dimensions.
CCSSM Mathematical Practices	4. **Model with mathematics:** Students create a model from which they can measure sides. 8. **Look for and express regularity in repeated reasoning:** Students observe and generalize the sequence of smaller squares which is a geometric sequence of lengths and areas.

Task

Supplies	Core Activity
• Sketch of nested squares	Students complete a drawing, organize their data using tables or lists, and support their conclusions.

Launch	Extension(s)
Use the launch activity to provide a review of the arithmetic of square roots—such as $(\sqrt{3})^2 = \sqrt{3} \cdot \sqrt{3} = 3$.	What happens if you start with a rectangle and create the midpoint shapes? Or three-dimensional objects, beginning with cubes?

The Shrinking Square

Launch

These two activities review area and radicals to launch this task and assess student readiness.

Part A

Given a 5 by 6 geoboard or geodot paper, comprised of 30 pegs, as shown, your task is to determine how many different sized squares can be created by connecting 4 pegs. Note that some squares, as shown below, will be "slanted" squares.

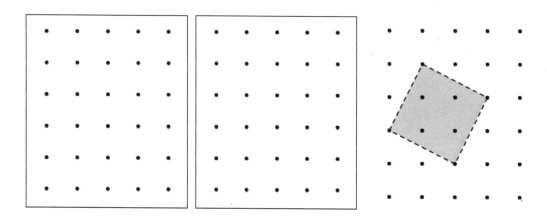

For each square you find, determine its area in square units.

Part B

Look at each row in the table below and explain whether or not the expression and the simplified expression are equivalent. Be sure to justify your answers.

Expression	Simplified Expression
$\sqrt{25}$	5
$\sqrt{7}^2$	7
$2\sqrt{3} \cdot 5\sqrt{3}$	30
$\sqrt{18}$	$3\sqrt{2}$
$\sqrt{6}^3$	$6\sqrt{6}$

The Shrinking Square

Activity

The design below started with an 8 unit x 8 unit square and was continued by constructing a series of smaller nested squares by connecting the midpoints of the sides of each square.

1. Continue this drawing for the next two smaller squares.

2. Construct a table showing the square number, the side length, and the area for the first six squares in this drawing.

3. If a side length of the original square is x, what is the side length of the nth square in this pattern? Justify your answer.

4. If the area of the original square is A, what is the area of the nth square in this pattern? Justify your answer.

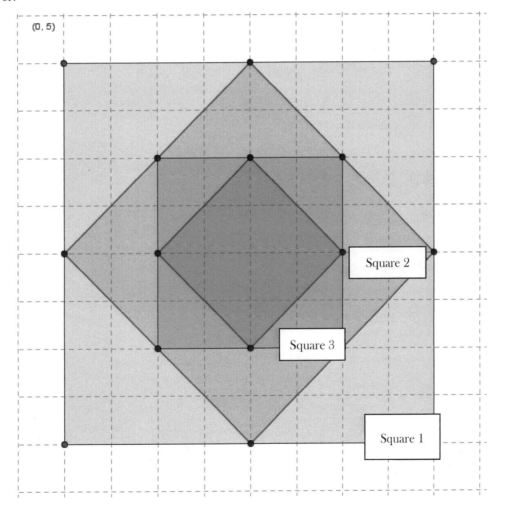

5. Now start with a rectangle that is 12 cm by 8 cm. Follow the same process of finding the midpoints and creating successively smaller quadrilaterals. Create a table of values for length, width, and area of each of these shapes, and generalize your results for the dimensions and the area. Then compare the findings for the rectangle with those of the square.

8th Grade   The Shrinking Square

The Shrinking Square

Results from the Classroom

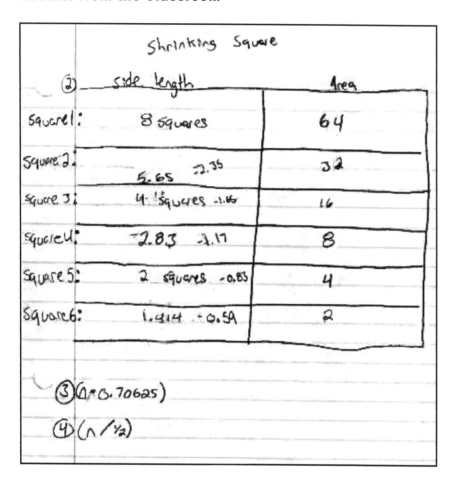

Two students worked on this task as a team and both of them showed the same work in their attempts at finding a pattern.

Having worked with finding the linear models for number patterns, you can see that they first tried to find the difference between the lengths of the sides. {-2.35, -1.65, -1.17, ...}

They were given a hint to look at ratios and headed off to actually find the pattern they wrote for part 3 ($n \cdot 0.70625$). That is fairly close to the actual pattern of dividing the previous term by the $\sqrt{2}$ or multiplying by $\frac{1}{\sqrt{2}}$.

Clearly neither student had connected the formula to the pattern that you get from multiplying by roots. Both exhibited the need for further work with simplification of radicals and the recognition or multiplication patterns with roots.

Problem 4 is an entertaining misconception that can even catch Calculus students. Ask most students to quickly take 60 divided by 1/2 and add 7, and you will get the wrong answer (37). We all know that they multiplied by 1/2 and did not divide by 1/2 since, in order to divide by 1/2, we have to multiply by the reciprocal 2.

Trouble with Trees

8th Grade—Teacher Notes

Overview	
Students will make connections between diameter and circumference of a circle.	**Prerequisite Understandings** • Define, evaluate, and compare functions.

Curriculum Content	
CCSSM Content Standards	8.F.4. Construct a function to model a linear relationship between two quantities. Determine the rate of change and initial value of the function from a description of a relationship or from two (x, y) values, including reading these from a table or from a graph. Interpret the rate of change and initial value of a linear function in terms of the situation it models, and in terms of its graph or a table of values. 8.F.5. Describe qualitatively the functional relationship between two quantities by analyzing a graph (e.g., where the function is increasing or decreasing, linear or nonlinear). Sketch a graph that exhibits the qualitative features of a function that has been described verbally.
CCSSM Mathematical Practices	4. **Model with mathematics**: Students are asked to apply a formula they have created to a real situation. 6. **Attend to precision**: Students need to calculate to a specified precision in the table they complete. 7. **Look for and make use of structure**: Students are asked to relate the calculations in their table to a graph.

Task	
Supplies • Chart paper and marker • Meter sticks, yardsticks, or rulers • Circular objects like bowls, cups, lids, etc.	**Core Activity** Students measure circular objects, complete a table and scatterplot, analyze their results, and answer a set of questions.
Launch Check student accuracy with measuring objects to the nearest centimeter. Attending to precision includes using appropriate vocabulary and recognizing relations that are and are not functions.	**Extension** Students could extend the relation to discover the rule for circumference as a function of radius.

Trouble with Trees

Launch

Fun with Functions

If needed, students should review the definition of a function and practice determining whether a relation is indeed a function. A short practice set like the one below will assess their readiness.

Which of the following are functions?

A. {(0,1), (-4,5), (7,5), (2,6), (3,5)}

B. {(-2,4), (5,5), (9,-9), (-2,-6), (1,0)}

E.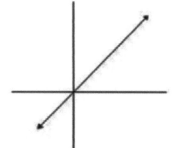

C.
x	y
2	-1
4	-3
6	-5
8	-7
10	-9

D.
x	y
1	-5
3	4
8	6
3	-4
1	5

Measuring Circumference

Students should practice measuring containers before beginning. Allowing students to work with smaller objects first, like a pill bottle, will ensure they understand how to roll the object along the measurement tool. Marking the start/end point on the object is a best practice. Students should work in groups of 2-3 and take turns measuring the objects and agreeing upon the measurement. Verify with each group they have correct measurements.

Activity Instructions

In this task, students will accurately measure circumferences and diameters of household containers, create an input-output table ("t-chart") of the measurements and graph the data, calculate the rate of change of the relation (find an approximate value for *pi*), and determine a rule for the relation (equation for circumference). They can also look for connections to slope, rate of change, linear equations, and/or proportions.

Additional Resources

Appendices A & B contain instructions for **Entering Data, Plotting Data,** and **Calculating Regression Equations** on both the TI-Nspire™ handhelds and the TI-84 Plus graphing calculators.

Trouble with Trees

Activity

Last spring, Mrs. Chang noticed one of her trees was not doing very well. After purchasing fertilizer, she read the directions and learned she needed to apply one pound of fertilizer per inch of tree diameter. Mrs. Chang wondered how she could accurately measure the tree's diameter precisely without cutting down the tree.

1. Complete the chart below by measuring five or more circular objects in centimeters. Then create a scatterplot of your results. Write your ratio as a decimal rounded to the nearest hundredth. When finished, average your ratios.

Circular Object	Circumference (C)	Diameter (d)	Ratio $\frac{C}{d}$
		Average	

8th Grade — Trouble with Trees

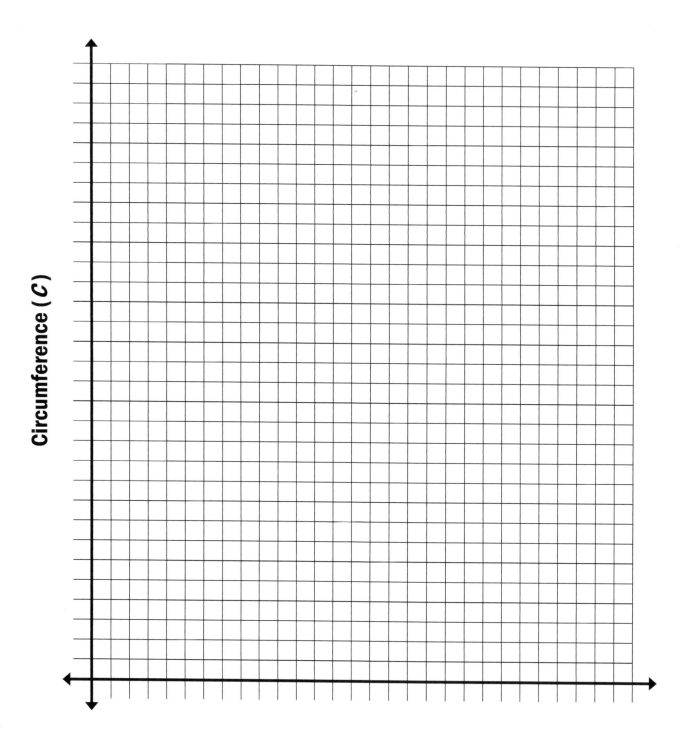

2. Record your observations about the data in the table and the resulting scatterplot.

3. Interpret the ratio of $\frac{C}{d}$ as a relationship between the circumference and diameter of circular objects.

4. How is the average ratio related to the graph?

5. How can circumference be used to find the diameter of a circular object?

6. Write an equation to represent circumference (C) as a function of diameter (d).

7. If the circumference of Mrs. Chang's tree measured 120 in., what would the diameter of the tree be? Show all of your work or explain how you found your answer.

8. Use the diameter you found to determine how many pounds of fertilizer Mrs. Chang should apply. Show all of your work, or explain how you found your answer.

Trouble with Trees

Trouble with Trees

Results from the Classroom

That is one big tree! This task is a nice confidence builder for students. Most students found it fairly straightforward, and none of them had much of a comment on the 120" tree. You might want to point out the importance of full sentences and communicating answers.

These two responses both offer a chance to see some variety in writing style.

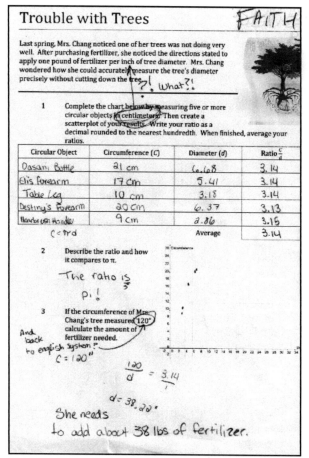

Faith

On the left you see the quick confident response that lets you know not only does Faith understand the problem but she is assertive in her answers. "The ratio is pi!" leaves no doubt in her conclusion.

The final answer which was listed at the bottom shows a minimal display of the proportion used to compute the answer.

The funny part is Faith's obvious consternation with the author of this Great Task. Why would the author dare to switch to centimeters in the middle and then back again to inches?

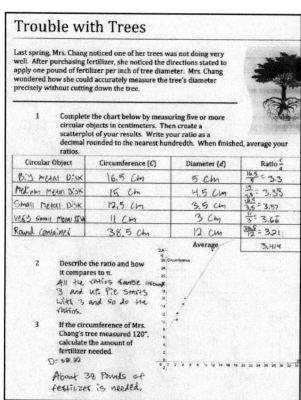

Other Students

A pair of students worked on the problem on the right. The group answered in full sentences and started to elaborate. Perhaps the space in the problem dictated the length of the answer?

As a follow up, you might ask, "Do they have trees like this in your neighborhood? That is a big tree."

Irrational Thinking, Inc.

Number and Quantity—Teacher Notes

Overview	
Student work with rational numbers to solve a series of challenges about operations with rational expressions.	**Prerequisite Understandings** • The basic properties of algebra. • The rules of simplification of rational numbers.

Curriculum Content	
CCSSM Content Standards	N.RN.1. Explain how the definition of the meaning of rational exponents follows from extending the properties of integer exponents to those values, allowing for a notation for radicals in terms of rational exponents. A.SSE.2. Use the structure of an expression to identify ways to rewrite it.
CCSSM Mathematical Practices	7. **Look for and make use of structure:** Students get to take a matching exercise and then are challenged to "prove" why the matches they choose really work using algebraic rules. 8. **Look for and express regularity with repeated reasoning:** Students develop an understanding of operations with rational numbers by looking at a large number of problems.

Task	
Supplies • None required.	**Core Activity** Students are engaged in testing suppositions and building knowledge of simplification rules.
Launch This activity is designed to refresh the students' understanding of perfect squares and simplifying rational numbers.	**Extension** Use similar methods to help students build rules for complex numbers.

Irrational Thinking, Inc.

Launch

Patterns, patterns, and more patterns – this is the stuff of mathematics! Learn to recognize the square numbers, and you are ready for an exploration into the way we work with numbers called square roots. In a crazy world that doesn't seem to act according to exact rules, these irrational numbers form the basis for opening your eyes to a whole new world of exact thinking!

1. Write down and learn the square numbers (x^2) from 1^2 to 25^2. The pattern starts with 1, 4, 9, 16, ... and continues to 625. Check your answers with your neighbors.

2. Extend and complete the below table of values for the exact simplified square roots of all the numbers from $\sqrt{1}$ to $\sqrt{100}$. Simplify by factoring out the largest perfect square factor greater than 1.

3. Explain why factoring out a perfect square of 1 doesn't simplify the results and why $\sqrt{30000}$ would simplify to $100\sqrt{3}$.

Root	Perfect Square Factor	Simplified
$\sqrt{1}$	$\sqrt{1}$	1
$\sqrt{2}$		$\sqrt{2}$
$\sqrt{3}$		$\sqrt{3}$
$\sqrt{4}$	$\sqrt{4}$	2
$\sqrt{5}$		$\sqrt{5}$
$\sqrt{6}$		$\sqrt{6}$
$\sqrt{7}$		$\sqrt{7}$
$\sqrt{8}$	$\sqrt{4 \cdot 2} = \sqrt{4} \cdot \sqrt{2}$	$2\sqrt{2}$
$\sqrt{9}$	$\sqrt{9}$	3
$\sqrt{10}$		$\sqrt{10}$
$\sqrt{11}$		$\sqrt{11}$
$\sqrt{12}$	$\sqrt{4 \cdot 3} = \sqrt{4} \cdot \sqrt{3}$	$2\sqrt{3}$
$\sqrt{13}$		
$\sqrt{14}$		
...		

Irrational Thinking, Inc.

Activity

Are you familiar with these four rules used for simplification of radical expressions?

Rule 1: Product Rule $\sqrt{x \cdot y} = \sqrt{x} \cdot \sqrt{y}$

Rule 2: Quotient Rule $\sqrt{\frac{x}{y}} = \frac{\sqrt{x}}{\sqrt{y}}$

Rule 3: Rationalize the Denominator Rule $\frac{x}{\sqrt{y}} = \frac{x \cdot \sqrt{y}}{\sqrt{y} \cdot \sqrt{y}} = \frac{x \cdot \sqrt{y}}{y}$

Rule 4: Simplify Perfect Squares $\sqrt{x^2 \cdot y} = x \cdot \sqrt{y}$

Patterns, patterns, and more patterns! You have been hired by Irrational Thinking, Inc., to use the rules above to match the problems below to the correct answer. Show how you can use the rules above to justify your thinking. Show the steps that get you to each answer!

	Problems		Answers
1	$\sqrt{24} \cdot \sqrt{6}$	A	$5 + 4\sqrt{2}$
2	$\sqrt{6300}$	B	$\frac{\sqrt{7}}{9}$
3	$\sqrt{12} + 5\sqrt{3}$	C	7
4	$\sqrt{\frac{7}{81}}$	D	$\frac{11\sqrt{3}}{3}$
5	$\frac{1 + \sqrt{7}}{2 - \sqrt{7}}$	E	$30\sqrt{7}$
6	$\frac{\sqrt{4900}}{\sqrt{100}}$	F	30
7	$\frac{11}{\sqrt{3}}$	G	$2 \cdot \sqrt[3]{7}$
8	$\sqrt[3]{8}$	H	12
9	$\sqrt[3]{27000}$	I	2
10	$\sqrt[3]{56}$	J	$-3 - \sqrt{7}$
11	$(1 + \sqrt{2}) \cdot (3 + \sqrt{2})$	K	$7\sqrt{3}$

Irrational Thinking, Inc.

Results from the Classroom

This task gives us pause to think about what makes a Great Task. Does a matching exercise where one is to show one's work qualify? In order to answer that question, look at three levels of students' responses.

Student #1

The first is an honors sophomore who clearly has some strong ideas on how to simplify rational expressions. She has no problem with getting the answers and can demonstrate how. For her, this activity was a review and perhaps helped remind her how to simplify while improving her proficiency by a spiral review of the topic.

Student #2

The second is a sophomore on grade level. This student shows her steps for most of the problems. Her work shows that she understands basic simplification skills and could use the task as a way of scaffolding the understanding to a new level. What can make this task a Great Task is the classroom discussion that is generated and the teacher's ability to create the atmosphere for extending skills to a higher level.

Student #3

The third sample of student work is telling in a couple of important ways. The student has some correct answers but no work. He has taken the time to write the problems but shows little of his thinking with steps.

The student work shows the need to find out more about what this student does know. Leading questions such as *"How could you demonstrate that number 1 is correct?"* could help the teacher discern the student's level of understanding. The student work might be a good opportunity for discussing the importance of both right answers and how to make those right answers understandable.

Stadium Dimensions

Algebra 1—Teacher Notes

Overview

Students will reason quantitatively with measurements and units and write equations to represent real-world situations with units.	**Prerequisite Understandings** • Units including inches and feet. • Writing equations with two variables.

Curriculum Content

CCSSM Content Standards	N.Q.1 – Use units as a way to understand problems and to guide the solution of multi-step problems; choose and interpret units consistently in formulas; choose and interpret the scale and the origin in graphs and data displays. N.Q.3 – Choose a level of accuracy appropriate to limitations on measurement when reporting quantities. A.CED.2 – Create equations in two or more variables to represent relationships between quantities; graph equations on coordinate axes with labels and scales.
CCSSM Mathematical Practices	3. **Construct viable arguments and critique the reasoning of others**: Students will 'sell' their design for a stadium. 4. **Model with mathematics**: Students will construct sketches of the design of the stadium. 6. **Attend to precision**: Students label all measurements and variables.

Launch Activity

Supplies	Core Activity Tasks
• Tape measures	Students will create stadium designs with units that make sense.
Introduction The launch activity gets students thinking about quantities, units, and reasoning.	**Extension(s)** Ask students how they would provide extra seating either with extra stadium rows or in an alternative location.

Stadium Dimensions

Introduction

Note: The introduction might take long enough that it needs to be assigned during the class period before or as homework to be discussed prior to the task.

The cheerleaders at Moo U (a Premier Agricultural and Engineering University) are going to make spirit flags for fans to wave around at the games. The flags will be a rectangle cut out of material and will have the school's name painted on it. The game is tonight, and the cheerleaders only have one piece of material that measures 54 inches by 10 feet.

1. Create a plan. Determine what size flags the cheerleaders should make to use all of the material and get as many flags as possible. How many flags can they make?

2. Why did you choose the size you did? How can you be sure the flags can be seen? Make sure to use units to explain.

A local rancher offered to pay for all the flags and all the material to make enough flags for 200 fans if the cheerleaders could help him solve the following problem.

Farmer Brown has 270 cows and wants to lease some acres to feed the cows knowing that the land he is looking at can feed 6 cows per acre for a period of 4 months and then the land will need to be cow free for 3 months to replenish itself. How many acres will he need to lease for a year to handle his cows?

3. Find a solution to this problem, and then clearly illustrate how you found the answers using dimensional analysis to explain each step in the solution.

Stadium Dimensions

Activity

The Moo U Board of Directors is planning to build a new football stadium. The seating capacity must accommodate a minimum 10,000 fans for the university to be eligible to host league events. The College Algebra class has been given the job of figuring out how big the stadium should be for this many fans to have seating.

In solving this problem, be sure to clearly explain your solutions for your design and how you determined the seating capacity of the football bleachers. Of course the Board of Directors also wants all units clearly labeled.

1. How much room should be allotted per person? What is this measurement in inches and in feet? How did you arrive at this figure? Is your measure based on inches / backside?

2. Sketch a drawing of the bleachers (both home and visitor sides) and include all measurements that are needed to calculate capacity.

3. How many people can sit on each row of the bleachers? Include calculations and units and what level of error you expect in the seating estimates.

4. How many total people can sit on the home side of the bleachers? How many people can sit on the visitor's side?

5. How many total people can sit on the visitor side of the bleachers? Show your calculations below, and explain how you rounded your answer. Remember to label units in your calculations.

6. Create a formula for the total capacity of the stadium (C) using x for the required length in *inches* to be left per seat. Use other variables as needed, and make sure to define the variables and the units used. Describe why you structured the equation or formula the way you did.

7. What other consideration for capacity would expect an architect to consider in the actual design of a stadium such as this?

Stadium Dimensions

Results from the Classroom

Did you know the average woman needs a width of 14.35" while the average man needs 17" for seating in a stadium? This is one of the

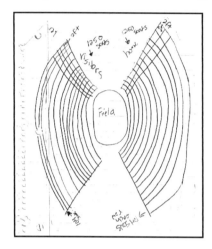

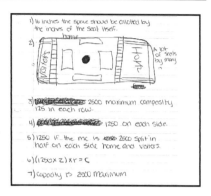

calculations students arrived at while creating a problem for designing a stadium. The work to the left and the right show some of the calculations and sketches students created to answer the questions in this task.

The students designed stadiums for 400 to 10,000 people. In general, their designs were simple rectangles and formulas they generated were all based on multiplicative formulas. The level of algebra demanded could be increased by introducing the challenge of estimating a cost for seating where there is a set $2/ linear feet with a 2% increase for each row you go up in height. You could also challenge them to produce some reserved seating with a premium of 50 for building each of those seats. Ask them to clearly delineate what is the cost of each type of seating. The sample work to the left illustrates how tasks are often met with short answers and simple formulas (Capacity = $x \cdot 1\frac{1}{2}$) where I assume the 1 ½ is linear feet per seat.

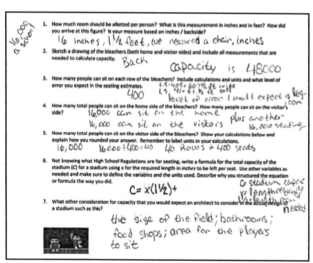

This task is a nice initial task for first-year students to introduce them to the need for labeling formulas and units and showing how they calculate dimensional measures. It also helps the teacher to define what directions can help students to explain their thinking.

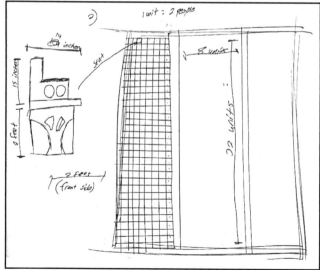

Students given this task had not been asked to solve a multistep task yet this year. This creates some cognitive dissonance. The task offered a good time to talk about the need for students to create their own assumptions. They learned to think about the capacity needed, shapes and parameters for drawings, and to accept the fact that not all problems have one right answer.

Numbers and Quantity

Chilling Out

Algebra—Teacher Notes

Overview	
This task asks students to explore equations with two or more variables.	**Prerequisite Understandings** • The difference between an expression and an equation. • Ability to analyze change in an expression with two or more variables.

Curriculum Content	
CCSSM Content Standards	A.CED.2. Create equations in two or more variables to represent relationships between quantities; graph equations on coordinate axes with labels and scales
CCSSM Mathematical Practices	5. **Use appropriate tools strategically**: Graphing calculators and a programming option are included. 7. **Look for and make use of structure**: Students examine inputs to formulas as parameters that they can control to examine how a multivariable function is used.

Task	
Supplies • Graphing paper • Graphing calculator • NWF Wind Chill .pdf file	**Core Activity** Students explore a common wind chill formula and evaluate the effect that certain variables have on the final temperature calculations.
Launch Students will learn to graph equations where one variable is kept constant and another one varies.	**Extension(s)** Take this project from mathematics into the realm of science. Formulas like $E = mc^2$ are classic icons in society and allow us to examine how different variables change the equation at different rates. Formulas are a nice way to introduce some of the ideas of calculus to students in earlier grades.

Chilling Out

Launch

Equations with more than one variable abound in mathematics. The most common is that of linear equations in the form $y = m \cdot x + b$. Below are equations describing 5 students' plans for saving some cash for college. The equation represents the money they have beginning on June 1 the summer before their freshman year in high school. Their total savings is represented by y while x represents the number of months that that they have been saving since June 1.

Student	Equation
Josiah	$y = 200 \cdot x + 550$
Julius	$y = 100 \cdot x + 1{,}000$
Juanita	$y = -250 \cdot x + 12{,}900$
Josephine	$y = 300 \cdot x$
Jasmine	$y = 200 \cdot x - 900$

1. Write descriptions to explain each person's savings plan.
2. How much money will each person have by June 1 of their year of graduation?
3. The rate of change in a linear function $f(x) = m \cdot x + b$ is m. Describe what the rate of change (slope) of these lines means in terms of the savings plan.
4. What could b (the y-intercept) represent for Jasmine?

Chilling Out

Activity

We have all felt the cooling effects of wind. A soft breeze on a warm day can help make the temperature more enjoyable and a strong wind on a cold day can chill us to the bone.

Below is the latest formula used by NOAA's National Weather Service for calculating how we perceive the temperature (T_{wc}=Wind Chill Temperature measured in °F) based on the actual temperature (T_a=Actual Temperature measured in °F) and wind speed (*V*=Velocity measured in miles per hour).[1]

$$T_{wc} = 35.74 + 0.6215 \cdot T_a - 35.75 \cdot V^{0.16} + 0.4274 \cdot V^{0.16}$$

1. Mentally compute T_{wc} when the temperature reads 0° F and the wind speed is 0 mph.

2. Create a table of values for a temperature of 50° F and wind speeds from 0 to 40 mph going by 5 mph each jump.

3. Create a table of values for temperatures from 0° to 40° with a wind speed of 50 mph.

4. Determine whether the wind speed or the ambient temperature has a greater effect on the wind chill. Justify your reasoning with data.

Extensions

1. Create wind chill graph for temperatures of 50° F.
2. Write a program to compute wind chill given the inputs of ambient temperature and wind speed.

---

[1] ***Note:*** *Wind chill temperature is only defined for temperatures at or below 50 degrees F and wind speeds above 3 mph. Bright sunshine might increase the wind chill temperature by 10 to 18 degrees F.*

Chilling Out

Results from the Classroom

Destiny

The work is straightforward from Destiny and reflects the fact that she is in an honors class and even feels comfortable with writing a program for the TI-83 Plus/TI-84 Plus graphing calculators.

She was also comfortable with her "illogical" answer to problem 1. Very few students were worried about that answer; most didn't realize that it made no sense. Occasionally a student would include a reference to the footnote in his or her answer as a justification of why the first calculation did not work.

The requested data tables were drawn on the back of the worksheet or on another sheet of paper and did not appear difficult for the students to complete.

Chilling Out

We have all felt the cooling effects of wind. A soft breeze on a warm day can help make the temperature more enjoyable and a strong wind on a cold day can chill us to the bone.

Below is the latest formula used by NOAA's National Weather Service for calculating how we perceive the temperature (T_{wc}=Wind Chill Temperature measured in °F) based on the actual temperature (T_a=Actual Temperature measured in °F) and wind speed (V=Velocity measured in miles per hour).[1]

$$T_{wc} = 35.74 + 0.6215 \cdot T_a - 35.75 \cdot V^{0.16} + 0.4274 \cdot V^{0.16}$$

1. Mentally compute T_{wc} when the temperature reads 0° F and the wind speed is 0 mph.
   35.74 $T_{wc}=0$ $T_a=0$ $V=0$
2. Create a table of values for a temperature of 50° F and wind speeds from 0 to 40 mph going by 5 mph each jump.
3. Create a table of values for temperatures from 0° to 40° with a wind speed of 50 mph.
4. Determine whether the wind speed or the ambient temperature has a greater effect on the wind chill. Justify your reasoning with data.
   Wind speed does because the range of the y's with different wind speeds range from about 66° to 3° while actual temperature ranges from -30° to -5.

Extra Home Projects for Experts

5. Create wind chill graph for temperatures of 50° F.
6. Write a program to computer wind chill given the inputs or ambient temperature and wind speed.

   ```
   Disp "Get wind chill"
   Disp "Give real temp"
   Input A
   Disp "Give wind speed"
   Input B
   Disp 35.74+.6215(A)-35.75(B)^.16+.4274(B)^.16
   Stop
   ```

[1] *Note: Windchill Temperature is only defined for temperatures at or below 50 degrees F and wind speeds above 3 mph. Bright sunshine may increase the wind chill temperature by 10 to 18 degrees F.*

Algebra · 88 · Chilling Out

Proving Patterns

Algebra—Teacher Notes

Overview	
Students will analyze quadratic patterns related to the difference of squares and use patterns with a number line.	**Prerequisite Understandings** • Multiplying a binomial by a binomial.

Curriculum Content	
CCSSM Content Standards	A.APR.1. Understand that polynomials form a system analogous to the integers, namely, they are closed under the operations of addition, subtraction, and multiplication; add, subtract, and multiply polynomials. A.SSE.2. Use the structure of an expression to identify ways to rewrite it.
CCSSM Mathematical Practices	2. **Reason abstractly and quantitatively**: Students move from abstract numerical calculations to proving them algebraically. 8. **Look for and express regularity with repeated reasoning**: Students develop the pattern for difference of squares.

Task	
Supplies • A number line • A table of squares	**Core Activity** Students will develop an intuitive formula for the difference of squares and then prove that relationship using the distributive property to multiply two binomials.
Launch Lead a class discussion on problem solving strategies such as make a list, make it table, look for patterns, and make a sketch. Ask students to list strategies that they have used in the past.	**Extension(s)** A discussion about the geometric view of what is happening could enhance understanding. If students rearrange 5 rows of 5, it is one more than 4 rows of 6. Try it!

Proving Patterns

Launch

Memorizing useful facts can help your ability to memorize overall, increase your understanding, make you faster at mental mathematics, and impress people. Performing memorization in two different locations can also make you remember better!

1. Take 10 minutes to learn the square numbers in the table to the left.
2. Cover the table, and have a neighbor quiz you.
3. Quiz your neighbor.
4. Memorize this at home and have someone quiz you.

What strategies did you use to memorize this table?

In the past, when you wanted to memorize something, what did you have to do to finally make that memory permanent in your mind?

N	n²	N	n²
1	1	13	169
2	4	14	196
3	9	15	225
4	16	16	256
5	25	17	289
6	36	18	324
7	49	19	361
8	64	20	400
9	81	21	441
10	100	22	484
11	121	23	529
12	144	24	576
		25	625

Extension

Geometric Justification

$$n^2 - 1 = (n+1)(n-1)$$

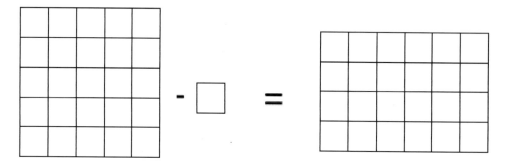

Algebra — Proving Patterns

Proving Patterns

Activity

Maria, who loves mental math, notices a strange pattern with the square numbers. She shared her findings with her friend, Ally. Maria asked Ally to select an integer on the number line. Then, Maria said, "Square that number. Then, take the number that is one greater and one less than your original number. Find the product of the larger and smaller values, and see what you get. It always happens that way!"

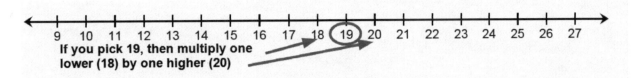

1. Select 5 values and then find the values of the product surrounding each value. Compare that value with the square of the selected number. What "always happens"?

Number					
Square					
Lower Number					
Higher Number					
Product of Lower·Higher					

2. Explain the pattern. Will this pattern work with negative numbers?

3. How does the pattern work with decimals and with fractions? (Provide examples to verify your answer.)

4. Let's prove that this pattern holds true for **any** value:

   a. If x is the original value selected, represent the square of the value.

   b. Represent the value one less than x.

   c. Represent the value one more than x.

   d. Find the product listed in the blanks from b and c above.

5. Write an equation that expresses the pattern that you explored and compare your answer with a partner.

Algebra — Proving Patterns

6. Take this pattern a step further. Select a value. Now, look at the values that are 2 units greater and two units less than the selected value.

   a. **Predict**: How will the square of the value differ from the surrounding two values?

   b. Verify your prediction by demonstrating the pattern using whole numbers.

7. Prove that the relationship holds true for any number.

8. What would happen if the values surrounding the given number were three units greater and three units less than the square of the selected value?

On the Flip Side

1. Write an expression for the area of the square, and an expression for the area of the rectangle.

2. This is an expression for the area of a different rectangle $x^2 - 16$. Write an expression of the product of the length and the width.

3. This is an expression for the area of a different rectangle, $x^2 - 49$. Write an expression of the product of the length and the width.

4. You have been writing the product of the length and width for several rectangles. Thinking about area enabled you to **factor** the algebraic expression. These expressions are all referred to as the "difference of squares." Why do you think that is the case? Explain your thinking.

5. Can you factor $x^2 + 9$? Why or why not?

Proving Patterns

Results from the Classroom

Lelah

Lelah changed the given table when she was solving the problems. This bright student must have known where she was going and made all the patterns into simple comparisons of the requested products to the values of n^2.

Her explanation in problem number 7 is short. She could be asked to frame the explanation in a sentence orally and ask her to reflect on how she could have written that.

Notice the quick error on problem 5.

Anthony

Anthony struggled more than Lelah. He filled out the table below, got lost in a difference pattern, and missed the connection. This is not an uncommon error when students are learning new pattern types.

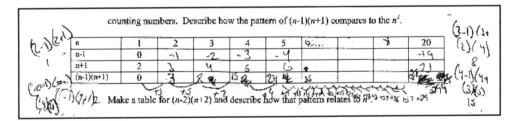

Waiting in the Queue

Algebra—Teacher Notes

Overview	
Students will explore the mathematics of waiting in line.	**Prerequisite Understandings** • None required.

Curriculum Content	
CCSSM Content Standards	A.CED.1. Create equations and inequalities in one variable and use them to solve problems.
CCSSM Mathematical Practices	1. **Make sense of problems and persevere in solving them:** Students engage in real sense making to relate the formula to the calculations. 2. **Reason abstractly and quantitatively:** Students explore the domain of a formula. 4. **Model with mathematics:** Look at a graph of a formula with parameters set.

Task	
Supplies • Graphing calculators	**Core Activity** Pairs of students use graphing calculators to investigate waiting in the lunch line at Normand High School.
Launch Students should be familiar with the concepts of domain and range of a function. Familiarity with interpreting functions and evaluating them for specific values is essential to this activity.	**Extension** Students collect data from their own cafeteria service at lunch. Students should come up with the average number of students in the queue and the average wait time.

Additional Resources

Appendices A & B contain instructions for **Entering Data** and **One-Variable Statistics** operations on both the TI-Nspire™ handhelds and the TI-84 Plus graphing calculator.

Waiting in the Queue

Launch

1. Consider the graph of $y = 2x^2 - x - 4$.

    - What is the domain of the function?
    - What is the range of the function?
    - What is the value of x when $y = 0$?

    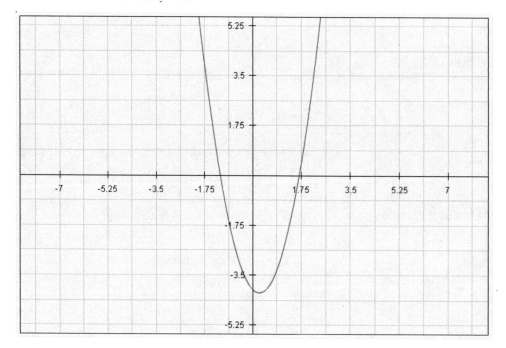

2. We use the formula $d = rt$ to represent distance, where r is rate and t is time. If you are driving 280 miles to visit your grandmother, what rate is necessary to make the trip in 4 hours? How much time is necessary if the average rate in miles per hour is between 50 and 65?

Waiting in the Queue

Activity

At Normand High School, students seem to spend a bit of time in the lunch line in the school cafeteria. Teacher Kelly Smothers wanted to know more about wait times, so she had students collect data over a week-long period. They found that the average number of customers arriving in an hour was 450, and the average number of students the servers can assist in an hour was 480.

1. Given that an average of 450 students arrives each hour to be served in the cafeteria, how much time occurs between successive arrivals?

2. Given that an average of 480 students can be served each hour, how long does it take for one student to get served lunch?

Kelly found a mathematical model for waiting in a queue. The model considers traffic intensity, represented by x, (which is the ratio of the *average rate of arrivals* to the *average rate of individuals being served*).

3. What is the traffic intensity for the lunch program at Normand High?

The formula further stated that the average number of students in the queue for lunch, L, can be represented by $L = \dfrac{x}{1-x}$

4. Using the data for Normand High School, find L.

5. What does this value of L tell you about the lunch service system at Normand High School?

Waiting a Bit Longer

Explore the waiting in line phenomenon by considering the graph of the function F (average time in line). On the graph, x is the ratio of *the average rate of student arrivals per hour* to *the average rate of students being served in the lunch line per hour.*

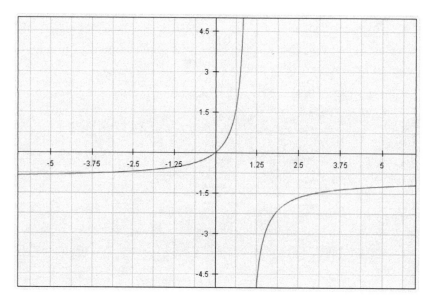

1. What is the domain of this function? What does this mean in terms of the context of this problem?

2. What is the value of L when $x = 1$?

3. What happens to L for $x > 1$? What does this mean in the context of this problem?

4. What does it mean for $x = 0$?

5. How could Normand High School improve cafeteria service wait time? How would improvements be reflected in the function L?

Waiting in the Queue

Results from the Classroom

The first part of the task asks students to find the time between successive arrivals and the time for one student to be served. These were easily identified by students as 1/(students each hour); and 1/(students served each hour): 1/450 hrs and 1/480 hrs. In the discussion, these were changed to minutes (.13.... minutes; and .125). Likewise, finding traffic intensity was not problematic. Students formed the ratio of the average rate of customer arrivals to the average rate of customers being helped: $\frac{450}{480}$ or $\frac{15}{16}$ = .9375.

Students moved quickly to apply the average number of students in the queue for lunch using the formula provided $L = \frac{x}{1-x}$ where x is the traffic intensity. Most students used the decimal value but a few groups used $\frac{15}{16}$. Students identified this as 15, but many did not indicate what this represented. When asked, some said *minutes* not *the average number of students in the queue*. Once this was cleared up, students offered "The lunch line appears to move fairly quick," "15 isn't bad if it is fast which appears to be the case given the time for one student to be served is .125 minutes," and "This seems fast. Maybe all students just go through and pick up a boxed lunch."

Another response worth considering was "15 in the line at one time makes it unreasonable that they could be served in .125 minutes. Of course, there could be multiple servers and limited choices. Maybe the queue is just for 1 item, and they get their drinks somewhere else or out a vending machine." The discussion provided opportunities for students to understand the variables that could be involved and the oversimplification of the problem.

Students were able to explore the graph of L with reporting out, providing an opportunity to interpret the information. Students identified the domain of the function quickly realizing that $x = 1$ would make the denominator equal 0. Students needed to be reminded that x represented the traffic intensity of (average rate of arrivals: average rate of individuals served). Once reminded, they were able to interpret that $x \neq 1$ because that would mean that arrivals were the same as the rate of being served.

Students realized that if $x > 1$ then the value of L would be negative and decreasing. Students realized that L is the average number of students in the queue for lunch, and you couldn't have a negative number of students.

Students felt Normand High School didn't need to improve their wait but added that larger values of x would result in larger values of L. Students also recognized that $0 < x < 1$ was the appropriate part of the function for this context.

Fractional Workers

Algebra—Teacher Notes

Overview

	Prerequisite Understandings
Students solve fractional work problems with pictures, graphs, and equations of one or two variables.	• Graphing linear equations. • Writing equations in one or two variables. • Solving additive fractional problems.

Curriculum Content

CCSSM Content Standards	A-CED.1. Create equations and inequalities in one variable and use them to solve problems. A-CED.2. Create equations in two or more variables to represent relationships between quantities; graph equations on coordinate axes with labels and scales.
CCSSM Mathematical Practices	4. **Construct viable arguments and critique the reasoning of others**: Students engage in rich discussions about the ways to interpret and model the concepts including using fractions and tables. 5. **Model with mathematics**: Students use pictorial, graphical, numerical, and algebraic models.

Task

Supplies	Core Activity
• Graphing paper • Graphing calculator	Students explore the working rate of three students and investigate three models (pictorial, graphical, and algebraic) that describe working situations.
Launch	**Extension(s)**
Students practice reading and thinking about fractions on a graph and then graph work done vs. time.	Students could look at using 2 equations with 2 unknowns. After the activity, students should discuss, identify and solve rate problems in the form $\frac{t}{2} + \frac{t}{4} + \frac{t}{3.5} = 1$ where the fractions represent the different speeds of the each worker.

Fractional Workers

Launch

Otto and Sparky are famous "Duck" artists. They were bragging in front of Mary the Mathematician about how fast they can paint a duck picture and how much detail they included. The next day Mary presented them with a graph of time vs. paintings done. Examine the pictures, Mary's graphs, and answer the following questions. Otto painted the two pictures on the left, and Sparky painted the two on the right.

Otto **Sparky**

Mary's Graph

1. Describe the graph.
2. Who paints fastest? How can you tell from the graph?
3. How fast is Otto?
4. How fast is Sparky?
5. Describe the axes.
6. Use the paintings to provide clues on why Sparky might be slower at finishing paintings.

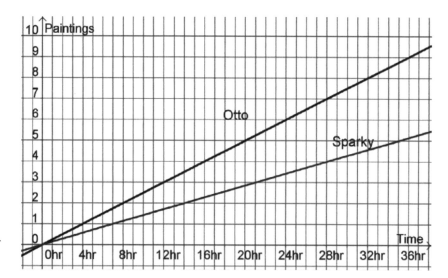

7. Write a function based on time for both people. p=f(t)
8. If Sparky and Otto work together to paint 23 pictures for a benefit auction, how long will it take?
9. Explain how you got the answer so that the benefit auction organizers will understand.

Fractional Workers

Activity

Sally, Sarah, and Suzie run the S^3 **Painting Service** where they paint old one-car garages for a summer job. They have lots of work because they live in an old part of town with lots of garages, and their fees are pretty cheap. They paint garages for less because they just love to fix up old garages so they look beautiful and elegant and make the world a better place. They also are saving lots of money for college!

Over the past few summers, they have figured out that speedy Sally can paint a garage in 2 days. Slower Sarah takes twice as long as Sally. Somehow Suzie can paint a garage by herself half a day faster than Sarah can paint one by herself.

Help us analyze this business.

1. How fast can each person paint a garage by themselves?

2. Make a sketch of 20 garages and show how much each girl will have painted after 7 days and how long will it take them to finish them all. Explain how you got your answer.

3. Create equations and make linear graphs that show how speedy Sally, slower Sarah, and somehow Suzie compare in their painting speeds. Let y represent the number of garages and x represent the time in days.

4. Figure how long would it take for Sally and Sarah to paint one garage together. Explain how you got your answer.

5. The S^3 girls have decided they can make a profit of $280 on each garage and that they will share the profits equally. Show how to figure out how much money they can make in a regular summer.

Fractional Workers

Results from the Classroom

Day	Sally	Sarah	Suzie
1			
2	✓		
3			
3 ½			✓
4	✓	✓	
5			
6	✓		
7			✓
8	✓	✓	
9			
10	✓		
10 ½			✓
11			
12	✓	✓	
13			
14	✓		✓

One of the students created a table of days with tally marks like the one at left. Hayley kept a tally of garages that were painted and just kept going until she got the correct number of garages. The table on the left illustrates the method, and you can see that by the 14th day that there were 14 garages painted. She was the first one to confidently get an exact day for her answer rather than a fractional day answer that she could justify.

Most students were fairly proficient is giving straight forward answers like the one below right. The one mistake made here came right at the beginning with assuming Suzie would take 4 ½ days vs. the actual 3 ½ from the description. This solution demonstrates how the first approach most students take is arithmetic.

It was expected that most students would use an algebraic solution like the one below, but it was rare to have it approached in the traditional manner.

$$\frac{t}{4} + \frac{t}{3.5} + \frac{t}{2} = 20$$

$$\frac{7t}{28} + \frac{8t}{28} + \frac{14t}{28} = 20$$

$$\frac{29t}{28} = 20$$

$$\therefore t \approx 19.3 \; days$$

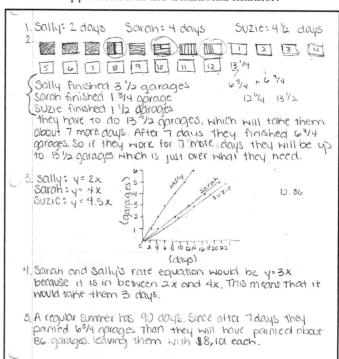

The students used a variety of approaches and could engage in the practices by sharing their different approaches. If none of the students use the algebraic solution in your classroom, it could an opportunity to share it with them and encourage them to compare it to the other approaches generated.

Dead Pennies

Algebra—Teacher Notes

Overview	
Students will use linear regressions to compute and model exponential relations.	**Prerequisite Understandings** • Scatter plots and line of best fit.

Curriculum Content	
CCSSM Content Standards	F.LE.2. Construct linear and exponential functions, including arithmetic and geometric sequences, given a graph, a description of a relationship, or two input-output pairs (include reading these from a table). A.CED.2. Create equations in two or more variables to represent relationships between quantities; graph equations on coordinate axes with labels and scales.
CCSSM Mathematical Practices	4. **Model with mathematics**: Students model an exponential decay with a fun penny-flipping activity. 5. **Use appropriate tools strategically**: Students use a calculator or program that can find an exponential regression. 6. **Attend to precision**: Students compare calculator results to the actual experimental probability generating a discussion about precision.

Task	
Supplies • Rolls of pennies • Graphing calculator or statistical software • Lid or flat surface	**Core Activity** Students simulate exponential decay by flipping pennies and removing those that land heads up. They will graph the data and analyze.
Launch Provide a review of linear regression. Use linear equations to predict a given outcome.	**Extension** Students explore their own examples of exponential decay.

Additional Resources

Appendices A & B contain instructions for **Entering Data**, **Plotting Data**, and **Calculating Regression Equations** on both the TI-Nspire™ handhelds and the TI-84 Plus graphing calculator.

Dead Pennies

Launch

Student	Missing assignments	Grades
1	3	86
2	0	84
3	9	67
4	16	68
5	33	50
6	1	79
7	2	69
8	3	81
9	20	60
10	17	63
11	13	66
12	17	63
13	13	71
14	34	41
15	23	53
16	15	45
17	28	56
18	16	62
19	15	69
20	19	65
21	2	84
22	1	97
23	4	76
24	11	57
25	9	59
26	47	18
27	11	76
28	11	68
29	9	70

The data at left is from an Algebra 2 Class first semester. There were a total of 68 assignments, including tests, quizzes, and extra credit. Graph the given data. Label the axes.

1. Describe the correlation as positive, negative, strong, or weak.
2. What observations can you make?
3. What can you say about the types of assignments that a student might have missing?
4. Enter the data into the graphing calculator. Enter Missing Assignments into L_1 and Grades into L_2. See calculator instructions if necessary.
5. Calculate the best fit line. Round to the nearest hundredth. What is the purpose of calculating this line?
6. The calculator gives you an r, which is the correlation coefficient. What is the correlation coefficient, and what does it tell us about our findings?
7. Predict your grade if you have only 1 missing assignment. No missing assignments?
8. If you want at least a B in Algebra 2, how many missing assignments could you have?
9. How accurate are our findings?
10. What are some conclusions you can make about this activity?

Build-in discussion time for the class to explain and support their conclusions. Note: Calculators will not generate an exponential decay with a 0 value for coins.

Extension

Be sure to check each of the student's sketches, before they get too far. Look for and give advice on how they can add up to 6 quadratic models. They will be approximating, so make sure to keep this in mind.

Algebra — Dead Pennies

Dead Pennies

Activity

Ollie the Ogre has passed out 50 pennies to you and your partner and asked you to play the game "Dead Pennies in Math Class." Here are the rules:

1. For round Zero, you have 50 pennies.
2. For the next round, you flip the pennies in a box and **take out any pennies that die**. A **dead penny** is one that lands heads up. Record the pennies that are still alive.
3. Continue with rule 2 with the just the pennies that are alive until all the pennies are dead.

Ollie the Ogre demands that you analyze his game, prepare a report so he can see if perhaps there is a fun way to market this to the masses and get rich! Analyze the graph by performing the following tasks.

1. Construct a table.

Toss #	0										
Pennies Remaining	50										

2. Graph the data, and describe the graph below.

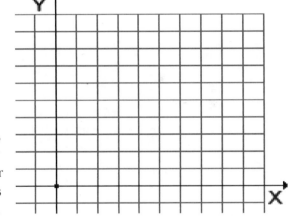

3. This graph looks like an exponential decay function, $y = a \cdot b^x$. How can you tell? What initial value for a would fit this function? What would you expect for a value for b? Write an exponential equation for this game based the theoretical probability of a coin toss.

4. Pick two actual data points. Find the exponential equation algebraically. $y = a(b)^x$

5. Pick two different data points. Find the exponential equation algebraically. Describe how close each equation is to the other. How can you tell these equations model exponential **decay**?

Algebra   Dead Pennies

6. Enter your data into a graphing calculator (Do not include the last toss where the remaining pennies are 0), plot the data in a scatterplot, and then calculate the exponential regression equation (which will not calculate a value if the last value for the pennies is 0).

7. Look at the precision that is suggested by the regression equation. How does this compare to your idea of a mathematical model for this problem?

8. Explain how your regression equation fits the rules of Dead Pennies in Math Class.

Extension

Do some research to find a situation and/or data that grow or decay exponentially. Use your book, the Internet, etc. Create your own experiment. Collect your own data. Show the data in tabular and graphic form. Then answer the questions below.

1. Give a short description of your research, experiment, and collection of data.

2. What did you learn? What did you find interesting?

3. Given your data above, find an exponential model algebraically.

4. Given the data, compute an exponential model using the graphing calculator.

5. Which model is more accurate? How accurate is it? Why or why not?

6. Propose a situation in which you could use this information. Be descriptive.

7. Summarize your findings, including any final thoughts.

Dead Pennies

Results from the Classroom

This Great Task fits in lots of categories. It clearly is algebra, could involve statistical modeling, and related to the function of exponential decay. It can also be used to emphasize to students the power and accuracy of a mathematical model.

Students should clearly understand exponential decay before doing this activity, perhaps with a problem in which they start with 100 ounces of gold and have to give 1/3 of it away every day. With true exponential decay, they will never have zero gold but should be quite close.

Note: When finding an exponential fit for this equation, calculators cannot fit the equation to data that have 0 values. If students are going to use technology, they have to eliminate the zero points. There is a real limitation to this model in this case because we cannot have 1/2 pennies. Discuss the limitation of models.

Once the students got the idea on how to get a close but not perfect model, then and only then, could we take the next step in talking about why a model is not a perfect fit for a situation.

The students were able to predict an average value for how many flips are needed to get to a zero penny result. The task lays some basic understandings that are important for students who will progress into statistics and probability.

The task was challenging for freshmen and sophomores. Part of the challenge is to put the ideas of algebra and functions together with a statistical situation that begs for a good mathematical model.

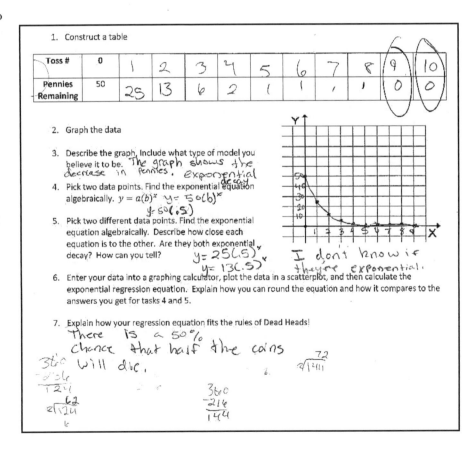

Algebra   Dead Pennies

Ferris Wheel Gala

Functions—Teacher Notes

Overview

	Prerequisite Understandings
Students are introduced to parametric functions using time as a variable and the positions of x and y as functions of time *t*.	• Transformations of linear, exponential, and quadratic functions. • How a geometric vector can be expressed as both an x- and a y-component. • How to generate a table of values for $x1(t) = 2t$ and $y1(t) = 3t$ in order to graph the points along the curve.

Curriculum Content

CCSSM Content Standards	F.IF.4. Interpret functions that arise in application in terms of the context.
CCSSM Mathematical Practices	1. **Make sense of problems and persevere in solving them:** Students will have to use some trials to set the parameters. 5. **Use appropriate tools strategically:** Students use a graphing calculator to test the trials. 6. **Attend to precision:** Students express the final answer in terms of *x*, *y*, and *t* solutions.

Task

Supplies	Core Activity
• Graphing calculators with parametric equation capabilities	Both sinusoidal and quadratic motions are used to introduce parametric functions.
Launch	**Extension(s)**
Students will translate the graphs of parametric equations to model circular and projectile motion.	A similar problem given a week later is a quick way to assess student level of mastery.

Additional Resources

Appendices A & B contain instructions for **Graphing Parametric Equations** on both the TI-Nspire™ handhelds and the TI-84 Plus graphing calculator.

Ferris Wheel Gala

Launch

Part A

1. Enter the following equations into a graphing calculator that is in parametric mode with the angle set to radians.

Problem 1

$x1(t) = 3(t)$
$y1(t) = 5(t)$

Problem 2

$x1(t) = 3 * \cos(t)$
$y1(t) = 5 * \sin(t)$

You should get a line and an ellipse. Discuss why the first function creates a line while the second creates and ellipse. How wide and how tall (major and minor axis) is this ellipse?

2. Vary the coefficients 3 and 5 in Problem 1 and describe how those values alter the graph.

3. Edit the original equations in Problem 2 as shown below. Explain what the -5 and +4 do to the graph.
   - $x1(t) = 3 * \cos(t) - 5$
   - $y1(t) = 3 * \sin(t) + 4$

4. Discuss and predict what equations for $x1(t)$ and $y1(t)$ that would create a circle with a radius of 12 centered at (5.10). Test the predictions by graphing.

Part B

A soccer kick of 80 feet per second at an angle of 65° can be modeled with two equations that describe the x- and y-components of the motion.

- $x1(t) = 80 \cdot \cos(65°) \cdot t$
- $y1(t) = 80 \cdot \sin(65°) \cdot t - 1/2 \cdot 32 \cdot t^2$

1. Graph this parametric equation, and estimate how high and how far the kick goes before it hits the ground. When is it at the highest, and when does it hit the ground?

2. What part of the equation controls the gravity?

3. What happens if you increase the velocity to 80 ft/sec?

4. What happens if you change the angle to 120°?

Ferris Wheel Gala

Activity

Two great applications of parametric equations are those of modeling the motion of a Ferris wheel and modeling the motion of projectiles. In this task, we combine both ideas.

Problem: Advise Ella as to the best angle and velocity to toss an apple to Tom.

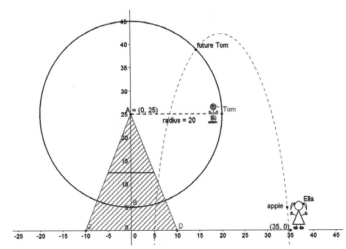

Tom is riding a Ferris wheel, and his friend Ella (seeing that he looks a bit hungry) decides to toss him a Gala apple. At time 0, Tom is exactly ½ way to the top of the 45' tall wheel.

Ella and Tom, both loving math, know that the wheel has a radius of 20', is 5' off the ground, and is rotating counterclockwise at the astounding rate of 6 *rpm* or 1 revolution every 10 seconds!

Ella is standing 35' away from the middle of the base of the Ferris wheel and will let the apple go at the exact height of 6' with the intent that it should arrive within 3' of long-armed Tom. Ella is allowed to move 5' closer or up to 10' further away. She can also build up suspense by waiting until Tom comes around again to make the toss!

1. Use parametric equations to model the motion of the Ferris wheel and the motion of the apple.

2. Describe how you came up with the equations. (If you used guess and check, tell where it was appropriate.)

3. Sketch a graph showing the path with the time and location of the toss and the time and location of the catch.

4. Check your solution on a graphing utility or graphing calculator using parametric equations. Stop the time at the moment of the catch to help justify your answer.

Ferris Wheel Gala

Results from the Classroom

Leah

Leah shows proficiency on this problem. She made a few smaller errors. Notice that she is missing a set of parenthesis on the y equation for the apple toss for the input for the cosine ($\frac{4\pi}{9}$).
Another problem comes in recording the angle. In her calculator, she had actually entered the correct angle of $\frac{5\pi}{9}$. You need the complimentary angle to get to the Ferris wheel if it is left of the toss.

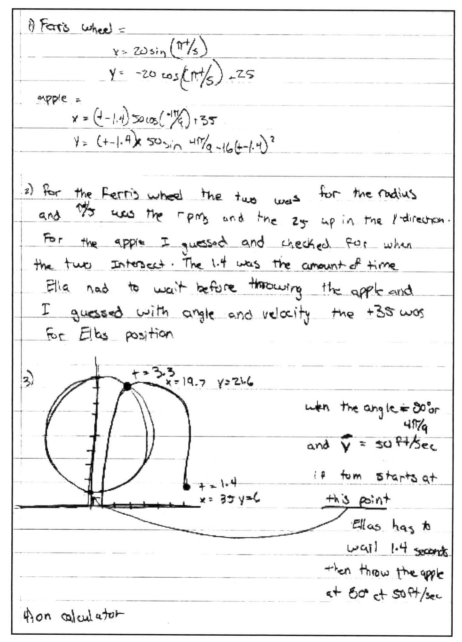

This problem is a good one for students to work on the practice of perseverance! They often have to use guess and check to get a toss that shows a nice soft arc rather than a "bullet" toss that has unlikely velocity.

Rise and Shine

Functions—Teacher Notes

Overview	
Students transform a sinusoidal function so that it models sunrise and sunset times for a specific location.	**Prerequisite Understandings** • How to change the amplitude, period, horizontal phase shift, and vertical phase shift for trigonometric functions.

Curriculum Content	
CCSSM Content Standards	F.TF.5. Choose trigonometric functions to model periodic phenomena with specified amplitude, frequency, and midline.
CCSSM Mathematical Practices	4. **Model with mathematics**: Students create a sinusoidal graph to fit data. 7. **Look for and make use of structure**: Students create a subset of data that represents a large set.

Task	
Supplies • Graphing calculator • Sunrise data. The Gaisma website can be a source. http://www.gaisma.com/en/ Sample data from this website for Helena, Montana are used.	**Core Activity** Students can decide how much of the data they will use to build a sine or cosine function that models the data. They should develop a strategy of selecting data that is periodic, shows the high and low values and allows them to build and test the model function to make sure it is fairly accurate.
Launch Review how to control the amplitude, period, horizontal phase shifts and vertical phase shifts in sinusoidal functions.	**Extension(s)** Perform the activity for sunrise, sunset, dusk, and dawn times for the students' location.

Additional Resources

Appendices A & B contain instructions for **Entering Data, Plotting Data,** and **Calculating Regression Equations** on both the TI-Nspire™ handhelds and the TI-84 Plus graphing calculator.

Rise and Shine

Launch

1. Record the amplitude, period, horizontal phase shift, and vertical phase shift of this function:
   $$y = 5(\sin(3\pi (x - 4))) - 6$$

2. Find an equation that will model the graph below.

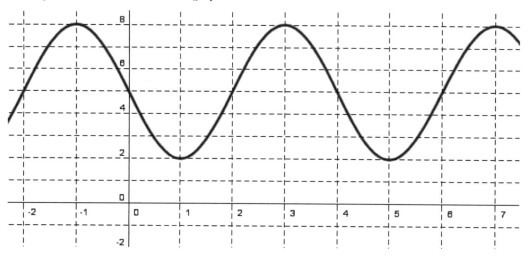

3. Describe how you change the period of a sine curve so than it repeats every 5 units.

Note: This activity assumes angles are measured in radians.

Rise and Shine

Activity

Sunrise time is a function of the time of year. In the United States, we are used to the fact that our sunrises are later in winter than in the summer. The farther north you travel, the more pronounced that effect becomes.

On the next page is data from US Naval Observatory in Washington, D.C. of the sunrise and sunset times in Helena, Montana.

Analyze the Data:

1. Create a scatter plot of sunrise times in Helena using a subset of this data that represents the whole year. Explain how and why you picked the subset of points for the graph.

2. Find a function that fits the data, and explain why you used the function you used.

3. Show a scatter plot with the function. Evaluate your model based on this graph.

4. Daylight savings time affects the local time of sunrise. How could you adjust your graph for the local daylight savings time?

HELENA, MONTANA
Rise and Set for the Sun for 2011

Location: W112 02, N46 35

Mountain Standard Time

Astronomical Applications Dept.
U. S. Naval Observatory
Washington, DC 20392-5420

Day	Jan. Rise	Jan. Set	Feb. Rise	Feb. Set	Mar. Rise	Mar. Set	Apr. Rise	Apr. Set	May Rise	May Set	June Rise	June Set	July Rise	July Set	Aug. Rise	Aug. Set	Sept. Rise	Sept. Set	Oct. Rise	Oct. Set	Nov. Rise	Nov. Set	Dec. Rise	Dec. Set
	h m	h m	h m	h m	h m	h m	h m	h m	h m	h m	h m	h m	h m	h m	h m	h m	h m	h m	h m	h m	h m	h m	h m	h m
01	0812	1651	0752	1732	0708	1814	0608	1857	0514	1937	0439	2014	0439	2025	0509	2000	0548	1908	0627	1808	0710	1713	0751	1643
02	0812	1652	0751	1733	0706	1815	0606	1858	0512	1939	0438	2015	0439	2025	0510	1958	0549	1906	0628	1806	0711	1712	0753	1642
03	0812	1653	0750	1735	0704	1817	0604	1900	0511	1940	0438	2015	0440	2024	0511	1957	0550	1904	0629	1804	0713	1710	0754	1642
04	0812	1654	0748	1736	0702	1818	0602	1901	0509	1941	0437	2016	0441	2024	0512	1955	0552	1902	0630	1802	0714	1709	0755	1641
05	0812	1655	0747	1738	0700	1820	0600	1902	0508	1943	0437	2017	0441	2024	0513	1954	0553	1900	0632	1800	0716	1707	0756	1641
06	0812	1656	0746	1739	0659	1821	0558	1904	0506	1944	0436	2018	0442	2023	0515	1953	0554	1858	0633	1759	0717	1706	0757	1641
07	0812	1657	0744	1741	0657	1822	0556	1905	0505	1945	0436	2018	0443	2023	0516	1951	0556	1856	0635	1757	0718	1705	0758	1641
08	0811	1659	0743	1742	0655	1824	0554	1906	0504	1947	0436	2019	0443	2023	0517	1950	0557	1854	0636	1755	0720	1703	0759	1641
09	0811	1700	0741	1744	0653	1826	0553	1908	0502	1948	0435	2020	0444	2022	0518	1948	0558	1852	0637	1753	0721	1701	0800	1641
10	0811	1701	0740	1746	0651	1827	0551	1909	0501	1949	0435	2020	0445	2021	0520	1946	0559	1850	0639	1751	0723	1701	0801	1641
11	0810	1702	0738	1747	0649	1828	0549	1910	0459	1950	0435	2021	0446	2021	0521	1945	0601	1848	0640	1749	0724	1659	0802	1641
12	0810	1703	0737	1749	0647	1829	0547	1912	0458	1952	0435	2022	0447	2020	0522	1943	0602	1846	0641	1747	0726	1658	0803	1641
13	0809	1705	0735	1750	0645	1831	0545	1913	0457	1953	0434	2022	0448	2019	0524	1942	0603	1844	0643	1745	0727	1657	0804	1641
14	0809	1706	0734	1752	0643	1832	0543	1915	0456	1954	0434	2023	0449	2019	0525	1940	0604	1842	0644	1743	0729	1656	0805	1641
15	0808	1707	0732	1753	0641	1834	0541	1916	0454	1955	0434	2023	0450	2018	0526	1938	0606	1840	0645	1742	0730	1655	0805	1641
16	0807	1709	0730	1755	0639	1835	0540	1917	0453	1957	0434	2024	0451	2017	0527	1937	0607	1838	0647	1740	0731	1654	0806	1641
17	0807	1710	0729	1756	0637	1836	0538	1919	0452	1958	0434	2024	0452	2016	0529	1935	0608	1836	0648	1738	0733	1653	0807	1642
18	0806	1711	0727	1758	0635	1838	0536	1920	0451	1959	0434	2024	0453	2015	0530	1933	0610	1834	0650	1736	0734	1652	0807	1642
19	0805	1713	0726	1759	0633	1839	0534	1921	0450	2000	0434	2025	0454	2014	0531	1931	0611	1832	0651	1734	0736	1651	0808	1642
20	0805	1714	0724	1801	0632	1841	0532	1923	0449	2001	0435	2025	0455	2014	0532	1930	0612	1830	0652	1733	0737	1650	0809	1643
21	0804	1716	0722	1802	0630	1842	0531	1924	0448	2002	0435	2025	0456	2013	0534	1928	0613	1828	0654	1731	0738	1649	0809	1643
22	0803	1717	0720	1804	0628	1843	0529	1925	0447	2004	0435	2025	0457	2011	0535	1926	0615	1826	0655	1729	0740	1648	0810	1644
23	0802	1718	0719	1805	0626	1845	0527	1927	0446	2005	0435	2025	0458	2010	0536	1924	0616	1824	0657	1728	0741	1647	0810	1644
24	0801	1720	0717	1806	0624	1846	0525	1928	0445	2006	0436	2025	0459	2009	0538	1922	0617	1822	0658	1726	0743	1647	0811	1645
25	0800	1721	0715	1808	0622	1847	0524	1929	0444	2007	0436	2025	0500	2008	0539	1921	0619	1820	0700	1724	0744	1646	0811	1645
26	0759	1723	0713	1809	0620	1849	0522	1931	0443	2008	0436	2026	0501	2007	0540	1919	0620	1818	0701	1723	0745	1645	0811	1646
27	0758	1724	0712	1811	0618	1850	0520	1932	0442	2009	0437	2025	0503	2006	0541	1917	0621	1816	0702	1721	0746	1645	0812	1647
28	0757	1726	0710	1812	0616	1851	0519	1933	0442	2010	0437	2025	0504	2005	0543	1915	0623	1814	0704	1719	0748	1644	0812	1648
29	0756	1727			0614	1853	0517	1935	0441	2011	0438	2025	0505	2003	0544	1913	0624	1812	0705	1718	0749	1644	0812	1648
30	0755	1729			0612	1854	0515	1936	0440	2012	0438	2025	0506	2002	0545	1911	0625	1810	0707	1716	0750	1643	0812	1649
31	0753	1730			0610	1856			0439	2013			0507	2001	0547	1909			0708	1715			0812	1650

Add one hour for daylight time, if and when in use.

Functions 118 Rise and Shine

Rise and Shine

Results from the Classroom

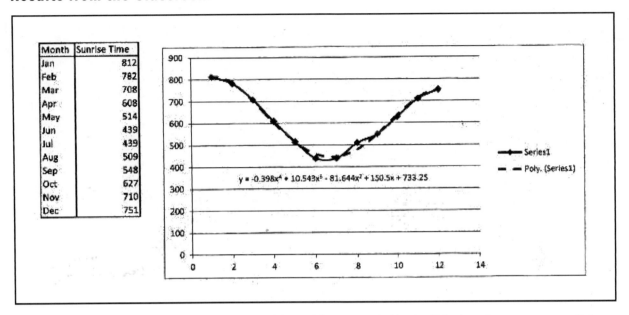

The student providing the sample answer for this problem was a junior in Calculus who was very proficient in solving traditional mathematics problems with technology but hampered by the fact that he didn't see the lack of justification and failure to explain what assumptions his model assumed.

In some ways, his approach was an elegant, quick-technology approach. He put the data for the 1st of every month in a table of values in Excel. Then, he graphed the data as a plot and finally he had Excel find a curve of best fit using a 4th degree model. A nice fit resulted!

Would this function give us a nice way to predict sunrise for any day of the year? Sure, but it might be important to use this example as a catalyst for the following discussions that might get the student to think about how a sinusoidal model would be a more informative model.

1. What other functions create this shape of a graph?
2. If you graph the sunrise times for 3 years in a row, what would the graph look like?
3. If the graph were a sinusoidal graph, what would be the amplitude and the period?
4. Can you use technology to create a sinusoidal graph, and compare the results to the 4th degree model above?
5. Did you explain your model, and was it clear to others what you did?
6. What might be a way to explain the model to make it clear to the average person on the street?

A Quadratic Smile

Functions—Teacher Notes

Overview	
This task is a review of quadratic functions and their properties and concludes with modeling to find a quadratic equation.	**Prerequisite Understandings** • Experience with quadratics, recognizing quadratic properties, and finding key features of a given model.

Curriculum Content	
CCSSM Content Standards	F.IF.4. For a function that models a relationship between two quantities, interpret key features of graphs and tables in terms of the quantities, and sketch graphs showing key features given a verbal description of the relationship. A.CED.1. Create equations and inequalities in one variable and use them to solve problems. A.CED.2. Create equations in two or more variables to represent relationships between quantities; graph equations on coordinate axes with labels and scales.
CCSSM Mathematical Practices	2. **Reason abstractly and quantitatively:** Students use photographs and diagrams to create equations. 4. **Model with mathematics:** Students connect curves to the quadratic equation model. 5. **Use appropriate tools strategically:** Students use calculators, graphs, and equations to describe information.

Task	
Supplies • Colored pencils or markers	**Core Activity** Start with students working alone. Move to small groups. Conclude with discussions and presentations of student work.
Launch Provide a review of quadratic functions. Students will understand the differences between parabolas with positive- or negative-leading coefficients.	**Extension(s)** Import real-world photographs into a graphing device. Find equations for the quadratic curves.

A Quadratic Smile

Launch

Scavenger Hunt

For each of the below classifications, list as many real-world objects that fit. Students could take a walk around school or search on the Internet. After individuals create a list, groups should compare and expand their lists.

1. List real-world examples of quadratic models with:
   - a minimum
   - a maximum
   - a wide width
   - a narrow width
   - orientated so that no minimum or maximum exists

2. Take your examples and discuss with your group how the general form of a quadratic would change to create the examples.

Finding Curves

3. Where do you see the quadratic? Trace any curve that you believe to be quadratic in red.

4. Sketch three curves that you believe to be quadratics, and use what you know to describe them.

A Quadratic Smile

Activity

1. Using the picture, trace only two quadratic models. Include one that opens up and one that opens down. Label them A and B.

For each of your curves:

2. Approximate the vertex, and label whether it is a maximum or a minimum.
3. Approximate the intercepts.
4. Write the equation of the axis of symmetry.
5. Use interval notation to indicate where the curve is increasing and where decreasing.
6. Describe the end behavior of each model.
7. Formulate an equation for each model.

Why would it be important to find the equation of a quadratic model? Describe a real life situation where this could be used.

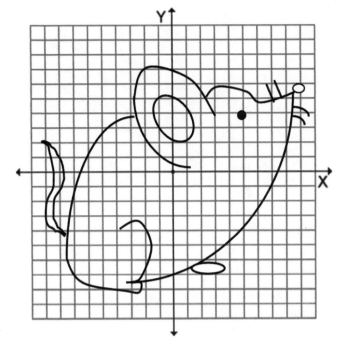

8. Now create a picture of your own. Sketch it on the graph provided below. Include at least 6 quadratic models using all 4 quadrants. Highlight the quadratic models, and label them 1 through 6.

For each model (1-6) find each of the following:

- Vertex
- Axis of symmetry
- Maximum or minimum
- x- and y-intercepts
- Increasing and decreasing intervals
- Quadratic equation

You will need to attach another sheet of paper to show work.

A Quadratic Smile

Results from the Classroom

Samantha

Sam's work here shows her understanding of each of the key features of a quadratic function, along with appropriate representation of ordered pairs, correct use of an equals sign, and precision in her approximations of the x- and y-intercepts.

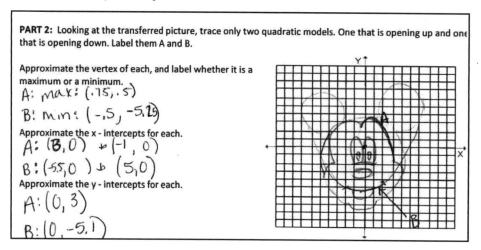

She forgot the brackets on A's interval notation for increasing and decreasing intervals, but on B she remembered it. This could be just a slight mistake, but she does seem to understand the concept. There could be good discussion on does the model actually go to + or - infinity?

On end behavior, she could use some help. She does notice that the quadratic models stop, so this could be used as a good discussion point.

Sam shows that she can create an equation while using appropriate tools strategically (the graphing calculator in this situation).

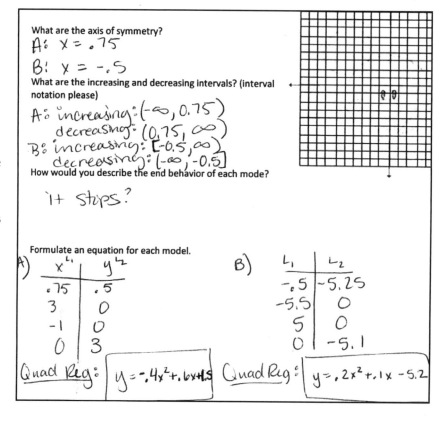

Jessica

In this extension activity, Jessica shows creativity and knowledge of quadratic visual representation as well as understanding of key features of quadratic models. She shows good use of an equals sign, interval notation, and estimation.

There could be valuable discussion in her increasing and decreasing intervals. She makes it a point to show that it does not go to infinity, but approximately stop at -6 and -3.

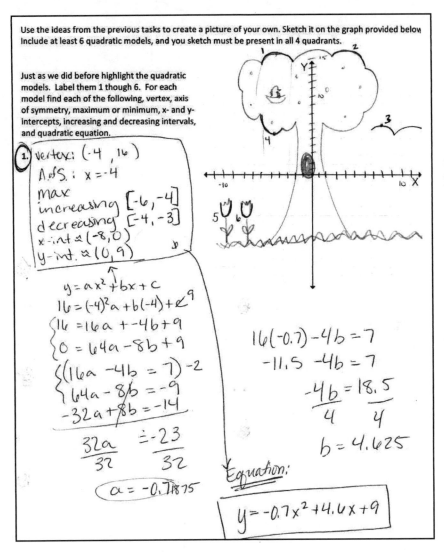

Jessica's work shows good understanding of creating a quadratic equation. She substitutes her known points into this formula, and then used elimination (linear combination) to solve her system of equations. There could be some discussion on how she substituted 9 into c. It could be a teaching moment to ask other students how she came to that conclusion.

Joe

In Joe's extension activity, there is good understanding of visual representation. There could be good discussion to distinguish if these are truly quadratic. Also, this visual includes a circle, an ellipse, straight lines, and a possible hyperbola (the ears.) This illustration could be used for a unit on conic sections.

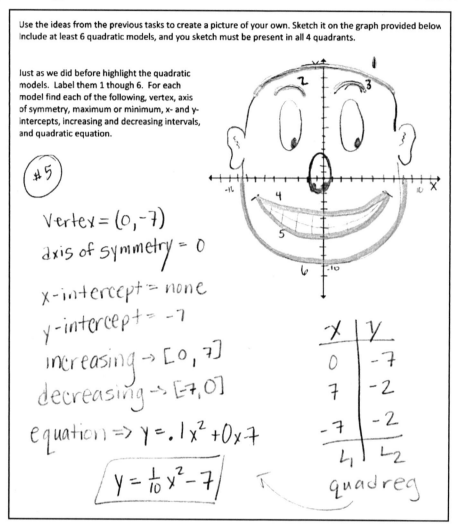

He shows some knowledge of the key features but has a misunderstanding of the use of an equals sign. He does have a good point that currently there is no *x*-intercept. Appropriate knowledge of interval notation, along with the appropriate use of tools strategically (the graphing calculator), is also demonstrated.

Joe shows precision in choosing the points (7, -2) and (-7, -2) which could be found on the actual line. This could be a good point to ask the students how he got those points and why it might be more appropriate to use these points and not the *x*- and *y*-intercepts that could be approximated.

Trout Pond

Functions—Teacher Notes

Overview

	Prerequisite Understandings
Students will use iteration, recursion, and algebra to model and analyze a changing fish population.	• Working with percentages and rate of change. • Predicting patterns in real-life situations with conditions and modeling data with tables.

Curriculum Content

CCSSM Content Standards	F.IF.4. For a function that models a relationship between two quantities, interpret key features of graphs and tables in terms of the quantities, and sketch graphs showing key features given a verbal description of the relationship. *Key features include: intercepts; intervals where the function is increasing, decreasing, positive, or negative; relative maximums and minimums; symmetries; end behavior; and periodicity.* F.BF.1. Write a function that describes a relationship between two quantities.
CCSSM Mathematical Practices	1. **Make sense of problems and persevere in solving them**: Students will generate tables by using recursive formulas. 4. **Model with mathematics**: Students use formulas in a variety of applications from biology. 7. **Look for and make use of structure**: Students illustrate a method that applies to solving many problems and is the basis for much of Excel's spreadsheet logic.

Task

Supplies	Core Activity
• None.	Students will organize their data in the table provided and discuss conjectures with each other.
Launch	**Extension**
Review calculating percentage amounts and rate of change.	Investigate how the fish population changes over time when one of the conditions is changed.

Additional Resources

Appendices A & B contain instructions for **Recursion Using a Previous Calculation**, **Generating a Sequence in a List**, and **Graphing a Sequence** on both the TI-Nspire™ handhelds and the TI-84 Plus.

Trout Pond

Launch

1. a. In a nature preserve, with a population of 2,000, the deer are disappearing at a rate of 15% of the current population every month. How many disappear the first month, how many are left, and how many disappear the next month?

1. b. A recursive formula for the number of deer is $a_n = .85 \cdot a_{n-1}$ where a_n represents the new population and a_{n-1} represents the previous month's population. Explain why this formula can be used for this situation.

2. An apple farm has 500 crates of red and green apples in its long-term warehouse. The farm expects to lose 10% of the total in storage to rot by the end of each month. Write a recursive formula for this situation, and make a table to show the number of crates of good apples left at the end of the next 4 months.

3. a. You're running a dog shelter for people to adopt dogs as pets. You started out with 200 dogs at the beginning of the year, and you adopted out 70% of the dogs during that year. Throughout the course of that year, you added 80 dogs back into the shelter. Explain the formula for this problem.

3. b. Many graphing calculators will do recursive formulas. Try this on a graphing calculator using the [ans] key:

   - Enter 200.
   - Then enter [ans] * .30 + 80.
   - Show the formula found a 70% decrease from the 200 and then added 80.
   - Repeat the calculation. Did it reduce the new total by 70% and then add in 80?

Trout Pond

Activity

Each spring, a trout pond is stocked with fish. Due to natural causes, the population decreases each year, and at the end of each year, more fish are added to the pond. Here's what you need to know about the number of fish currently living in the pond this year:

- There are currently 3000 trout in the pond.
- Due to fishing, natural death, and other causes, the population decreases by 20% each year, regardless of restocking.
- At the end of each year, 1000 trout are added to the pond.

1. Without calculating, predict the long-term population under these conditions. Will it grow slowly, grow quickly, grow without bound, level off, oscillate, or die out? Explain why you expect this result.

2. Make a table of values, like the one shown below to calculate the population at the end of the next 20 years. How does this compare to your prediction?

Year	Number of Trout in Pond	Year	Number of Trout in Pond
1	3000	11	
2		12	
3		13	
4		14	
5		15	
6		16	
7		17	
8		18	
9		19	
10		20	

3. Predict the population after 30 years. Then, extend your table, and compare the result to your prediction.

4. Predict and then calculate the long-term population of mountain lions if we started out with 200 lions in January, with a death rate of 20% by the end of the year and a birth rate of 1 cub for every 2 lions alive in January.

5. Create a recursive formula. Check the results on a calculator.

6. Create a graph of the long-term behavior of this function.

7. Discuss the model. What limitations do these models have? How could you make these models more realistic for real populations? How could you introduce more risk into the population growth?

Trout Pond

Results from the Classroom

The students' work on this problem was fairly straightforward and a bit enlightening! Some students wanted to use Excel for the task and went to the library to work in pairs. They came back by the end of the period with some but not the entire task completed.

Some of the best results came from students who used basic recursion on a graphing calculator. The example below was typical of students who did this. This one was the neatest. However, the second part dealing with the lion population should show exponential growth. Their formula must not have computed the birth rate correctly.

Some students added the 1,000 fish right away. They assumed the current date is right before the end of the year. Those different results can lead to a great discussion.

For extensions to this problem, have students create some variable rates to growth and add some probabilistic crashes. It would be interesting to see how they would try to vary rates each year from a 20 to 50% decrease with a 5% probability of disease wiping out 85 percent of the population every 20 years.

The debates about carrying capacity and unchecked exponential growth could be a powerful lead-in to **logistic curves** - an important idea in higher mathematics.

Fractal Doodles

Functions—Teacher Notes

Overview

This task takes a popular video topic (mathematical doodling) and goes further to explore the sequences of measures associated with fractal doodles.

Prerequisite Understandings

- Ability to examine a pattern of geometric growth with both numeric and algebraic representations.
- Ability to discern the difference between recursive and algebraic formulas for the nth term.

Curriculum Content

CCSSM Content Standards	F.IF.4. For a function that models a relationship between two quantities, interpret key features of graphs and tables in terms of the quantities, and sketch graphs showing key features given a verbal description of the relationship. A.CED.2. Create equations in two or more variables to represent relationships between quantities; graph equations on coordinate axes with labels and scales.
CCSSM Mathematical Practices	1. **Make sense of problems and persevere in solving them:** Students develop and build the patterns in tables. 7. **Look for and make use of structure:** Students find structure in the pattern in the tables. 8. **Look for and express regularity with repeated reasoning:** Students explore sequences of measures and look for relationships.

Task

Supplies	Core Activity
- Interactive geometry software or drawing tools - Graphing calculators or software (optional)	These mathematical "doodles" explore the ideas of recursive patterns. The activity gives students a chance to practice making some recursive sketches and then engages them in building tables and finding patterns of growth.
Launch	**Extension**
This activity is designed to increase student capabilities on finding the formulas for arithmetic and geometric sequences and remind them how to determine the arithmetic, geometric, and quadratic formulas.	Students with a good understanding of Java (or other appropriate programming language) might want to take the challenge of building the structure to create computer graphics.

Fractal Doodles

Launch

Patterns, patterns, and more patterns – this is the stuff of mathematics! Remind yourself how to find the formulas for arithmetic, Fibonacci, quadratic, and geometric growth. You should be able to find the pattern, list a few more terms, and present a formula for the nth term. Recall that there are recursive, or algebraic, formulas that can be used for patterns. Note: Fibonacci type patterns only have a recursive form of an equation [$a(n) = a(n-2) + a(n-1)$].

Find the pattern and formula for the following sequence of numbers:

Arithmetic Sequences

1. 5, 13, 21, 29, ...
2. 10 ½, 9, 8 ½, 7, ...

Geometric Sequences

3. 3, 6, 12, 24, ...
4. 2, 2.2, 2.42, 2.662, ...

Quadratic

5. -4, -6, -6, -4, 0, ...

Fibonacci

6. 1, 1, 2, 3, 5, 8, ...
7. ½, ½, 1, 1 ½, 2 ½, 4, ...

Drawing

8. △, □, ⬠, ⬡, ...

Extension

This project can be used as a starting point for students who want to take a project from mathematics into the programming class. Below is an example of a computer graphic of the fractal along with a Logo program to create it. This program was written in Terrapin Logo http://www.terrapinlogo.com/ .

```
TO TREE :DISTANCE
        IF :DISTANCE < 5 [STOP]
        FORWARD :DISTANCE
        RIGHT 30
        TREE :DISTANCE - 10
        LEFT 60
        TREE :DISTANCE - 10
        RIGHT 30
        BACK :DISTANCE
END
```

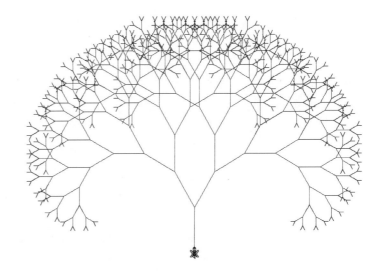

Fractal Doodles

Activity

Fractal Doodle 1: The Right Maze

A popular fractal doodle is shown below. Stage 0 is a segment of length x. Stage 1 adds two perpendicular segments on the ends that are 3/4 the original segment. The pattern continues with adding of new segments on the ends of the last additions. The new segments are always 3/4 the length of the previous stage's smallest segments.

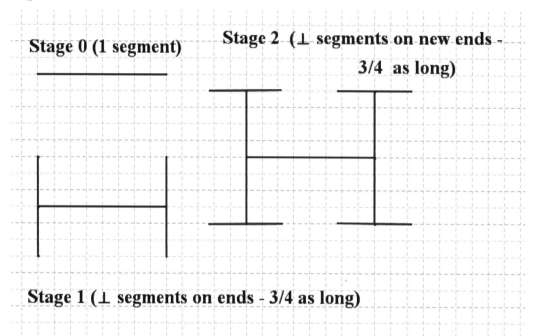

Use this pattern to complete the following tasks:

1. Create fractal sketches that go from Stage 0 to Stage 6.

2. Create tables and find the formulas for the patterns including:

    a. The lengths of the successive new segments.

    b. The total length of segments for each stage.

    c. The number of segments added at each stage.

    d. The total lengths of all segments.

    e. The average length of all segments.

3. If this fractal represents a maze and you enter one end, what is the probability you will randomly come out at each of the other ends of the segments for a Stage 5 fractal?

4. Is there a limit to this sequence of lengths as you to infinity? If so, explain how you would find it, and then find it.

Fractal Doodle 2: The Recursive Tree

Another exciting fractal is that of the recursive tree. Stage 0 is the trunk of the tree. At the end of that stage you add two smaller branches to create stage 1. Stage 2 adds more branches to ends of the Stage 1 branches. The angles of turns for the branches can be 30° left and right of the original branch like the picture on the far right or other angle combinations to create an asymmetric tree like those on the left.

Assume that you start with a main branch that is 2 m (2000 mm) tall.

1. If each successive branch is 20 cm shorter than the previous branch, explain the pattern of branch lengths, and explain the total length of all the branches in the tree.

2. If you are given that each successive branch is 90% of the length of the previous branch, make a table. Describe the growth pattern.

Extra Challenges

- If you are given the rule that all the **right** branches are 90% of the length of their previous branch, and all the **left** branches at 80% of the previous branch, make a new table of growth values. Describe the pattern.

- Can you create a computer program or script to create a fractal picture?

- Research fractals on the Internet.

Fractal Doodles

Results from the Classroom

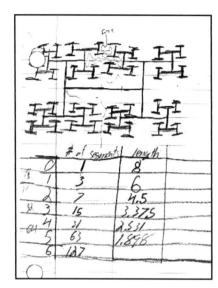

The student work here contrasts the work of a sophomore in an Honors Pre-Calculus class compared to two sophomores in a Geometry class.

The Geometry students had roughly 35 minutes to tackle the problem with a partner. The students show some promise with understanding the pattern but lack the mathematical sophistication to carry out the work to a formula. Up to this time, they had only looked at linear and quadratic formulas, such as the number of diagonals in a square, and neither partner was able to make the jump to an exponential formula.

The next student was able to get a much more complete set of data to include in the tables. She was also able to come up with a formula for the growth for parts a, b, and c. The level of ability to mathematize the problem and create models is clearly stronger with a student who has had 2 more years of mathematics.

The second student was still challenged (and unable) to come up with a formula for the total and average length of all segments but is clearly emerging in those skills. She must have had an introduction to sigma notation the year before and had a sense that this is where the formula had to go. It would be interesting to see how she would do at revisiting the problem after this year's work on sequences and series.

The Tipi

Geometry—Teacher Notes

Overview	
Students will work with dimensions, area, and volume of pyramids and prisms as they investigate the mathematics behind the design of a Western Plains Indian Tipi.	**Prerequisite Understandings** • Area, volume and surface area. • Calculate slant heights of pyramids and prisms using the Pythagorean Theorem or trigonometry.

Curriculum Content	
CCSSM Content Standards	G.MG.1. Use geometric shapes, their measures, and their properties to describe objects. G.MG.3. Apply geometric methods to solve design problems. G.GMD.3. Use volume formulas for cylinders, pyramids, cones, and spheres to solve problems.
CCSSM Mathematical Practices	4. **Model with mathematics**: Students create a polygonal pyramid model (or cone) for a tipi. 6. **Attend to precision**: Students' construction of the model tests precision in scaling down a real-life shape.

Task	
Supplies • Stiff paper for making models (optional)	**Core Activity** The emphasis in this activity is leading students toward exploring their own thinking, rather than the teacher leading them to any particular solution.
Launch Review and strengthen student skills on the Pythagorean Theorem, area formulas, volume formulas, and trigonometry with triangles.	**Extension(s)** Explore the traditional size of a tipi from research with old pictures, using people's heights as a determinate of the scale of a photo. Explore the usable floor area for standing or sitting in a tipi, find the maximum volume for a tipi with a given base and surface area, or compute the number of buffalo hides that were needed to cover a tipi.

The Tipi

Launch

Use geometric and trigonometric formulas to analyze the following geometric shape.

1. Compare the slant height and the volume of a cone with a height of 20 cm and a diameter of the base of 8 cm to the slant height and the volume of a square pyramid with a height of 20 cm and a base diagonal of 8 cm.

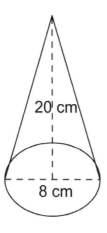
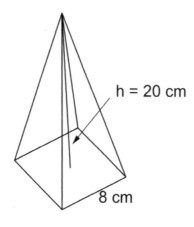

2. Find the altitude, missing side length, missing angles' measurements, and area of the triangle below.

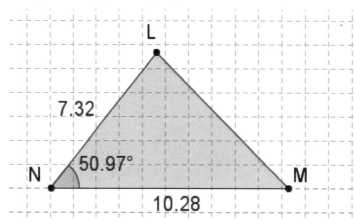

3. Describe how to find the surface area and volume of pyramids with a regular polygon as a base.

The Tipi

Activity

A strong part of western American culture is the Plains Indian Tipi. This structure could be as simple as a 3 pole structure (a tetrahedron shape) but usually was much more elaborate, with 12 or more poles giving the structures a cone shaped appearance.

Your task will be to create an engineering design and a **scale model of a life-sized tipi**. Provide specifications, and describe the processes you used to come up with the measurements. The life-sized tipi you will be modeling is a 12-pole tipi with a height of 18 feet to the cross pole, and has a perimeter of 48 feet (4 feet between poles).

1. Prepare a detailed specification sheet. It should include the following:
   - A title page with your name.
   - A drawing of the area of the base of the tipi with measurements of lengths and areas.
   - A drawing of the net (the cover of the tipi) with area calculations.
   - The calculated volume of the tipi.
   - A supplemental page where you outline the formulas and calculations you did to find the information above.

2. Create a scale model of the tipi.
   - Make your model from any materials you wish.
   - Make a title card to place with the model that includes its title, your name, and the scale of the model.

3. Decorate the cover with some traditional Plains Indian designs or geometric patterns.

The Tipi

Results from the Classroom

This task was tackled by a class that had just started the year with trigonometry and had an easy fluency with formulas and angles.

Fewer than 20% of the students chose the option of using a computer program to build their models. One student's computer-generated scale model of a side of the Tipi (the base, one side, and a side view) is shown on the right.

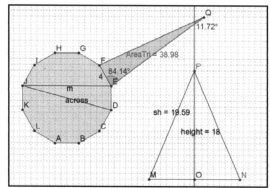

Most students chose to make all drawings and all calculations by hand on cm grid paper. The actual scale models turned in by most students used a scale of 1cm = 2 ft. It was clear that the size of the paper and open discussions during the project design and building phase resulted in students sharing their ideas for scale with each other. Most scale models came out very close in size because of this similarity of scale. To get more variety in the final results, have students use different scales or vary the heights and number of poles.

The project did result in some interesting errors by students. Take the list of formulas typed on one paper (see right). This student assumed the shape was just a cone and made some other basic errors such as the formula for circumference and the slant surface area (1/2 of a circle?).

Cone:
$V = (1/3)\pi r^2 h$
Circle/Base:
Perimeter$=\pi d^2$
$A = \pi r^2$
Net:
$A = \pi r^2 / 2$

The student work to the far right made extensive use of the Pythagorean Theorem and did not rely on trigonometric calculations for most angles in the drawings.

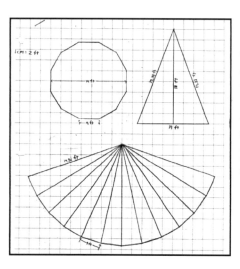

Many of the drawings looked like the one on the left. They were accurate, neat, readable, and involved the correct scale. Labeling all angles in the drawings would help reinforce the applications of trigonometry to this project.

As a final note, the real tipis of the America Plains Indians were actually made of animal skins and were stitched together in patterns that could be adjusted as they were being built. Trial and error along with experience certainly has a value!

Infinity Pizza

Geometry—Teacher Notes

Overview	
Students investigate slices of triangles to learn how midsegments and other divisions of a triangle affect the area.	**Prerequisite Understandings** • Find the area of a triangle both with the formula $A = \frac{b \cdot h}{2}$ and with Heron's formula. • Basic concepts of midsegments and formulas for distances between points.

Curriculum Content	
CCSSM Content Standards	G.GPE.4. Use coordinates to prove simple geometric theorems algebraically. G.CO.10. Prove theorems about triangles. Theorems include: vertical angles are congruent; when a transversal crosses parallel lines, alternate interior angles are congruent and corresponding angles are congruent; points on a perpendicular bisector of a line segment are exactly those equidistant from the segment's endpoints.
CCSSM Mathematical Practices	2. **Reason abstractly and quantitatively**: Students reason based on formulas or on coordinates. 3. **Construct viable arguments and critique the reasoning of others**: Students explain and justify their reasoning in groups before sharing with the class. 8. **Look for and express regularity with repeated reasoning**: Students explore how changing inputs will change formulas.

Task	
Supplies • None	**Core Activity** Give students time to discuss, question, think, investigate, try, and revise their work. Set clear expectations for descriptions of thinking and logic.
Launch Students will think about triangular areas and their relationship to the length of sides, the height and base of various triangles.	**Extension(s)** Square, rectangular, trapezoidal, and other shapes of pizzas might be fun extensions. Students could explore, create, bake, and bring to school to eat!

Infinity Pizza

Launch

1. Find the areas of the following triangles using the traditional area formula, Heron's formula, and trigonometry.

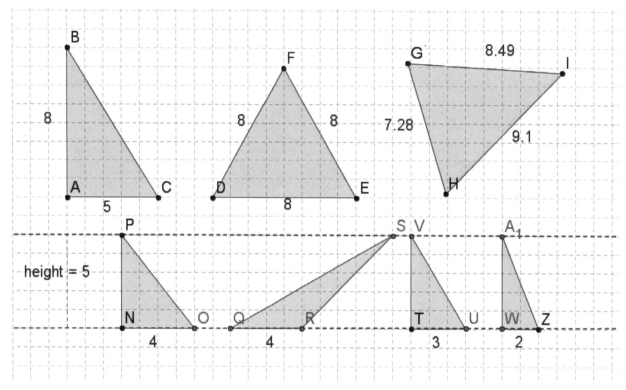

2. How does changing the bases in the last problem to 3/4 or 1/2 of the original bases affect the areas?

3. If you cut the base in a triangle down to 1/3 of its original size while keeping the height the same, what will happen to the area?

4. If you cut the base in 1/2 and the height in 3/4, what will happen to the area of a triangle?

Infinity Pizza

Activity

Exploration 1: Sally's Pizza Chop Shop specializes in gigantic pizzas made in random triangle shapes. You and two good friends decide to get a pizza and share it equally.

1. Devise a fair method for cutting any triangular pizza into 3 equal-sized pieces. Justify that your method creates a fair division using the formulas and properties of area for geometry.

2. Use an interactive geometry program or graphing calculator to model the problem and show that the areas are the same with that model.

Exploration 2: Sally uses her famous infinity cut to share a pizza with 3 people!

She first cuts along the lines from 1/3 of one side to 2/3 of the adjoining side. She claims that this gives each person an equal share (1/3 ∘ 2/3 = 2/9) and leaves a triangular piece (1/3 of the pizza) for a second cut.

Then she cuts the remaining triangle in the same manner, leaving 1/3 of that triangle. The cutting goes on and on using that same pattern.

Sally exclaims, "Since the pieces get smaller and smaller, you get to eat less and less making it easier to finish her giant pizzas!" The cut method is illustrated below.

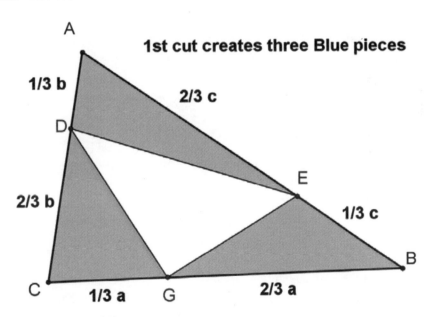

1. Explain why Sally's cut creates 3 equal shares.

2. Create a sequence of fractions that represent each person's share with each successive cut.

3. Does the sum of the sequence for each person's slices converge to 1/3? Explain.

4. Discuss the strengths and limitations to this method of pizza cutting and eating.

Infinity Pizza

Results from the Classroom

Exploration 1

The activity requires students to have a strong sense of how changing the base or height of a triangle affects the area. When given the first task of creating a strategy for cutting a triangular pizza into three equal shares, the most common response for a Geometry class just starting area is shown below.

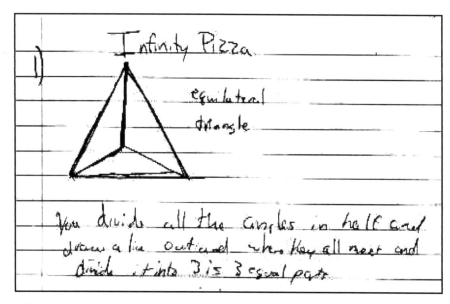

An easy solution is to start with an equilateral triangle, bisect the angles, and (Voila!) problem solved.

Students found this nice simple solution of choosing an equilateral triangle also worked if they constructed the medians, or the altitudes, or the perpendicular bisectors. It appears that work on circumcenters, incenters, centroids, and orthocenters all had made some impression on the students.

The question "What about scalene triangles?" sent them back to the drawing board, and it appeared that they didn't have strong area strategies for dealing with this problem without some extensive work on how to divide segments into equal parts.

Some pre-activities involving dividing squares, rectangles, and then circles can help give them a sense on how changing a factor in a formula alters the final answer. Once they build a surer sense of how bases and altitudes determine the size of the area of a region, then they should have a better sense of how to solve and explain the problem.

Exploration 2

As to the infinity cuts, a Pre-Calculus student upon seeing the problem mentioned that it might be easier to use medians since that gives you four equal pieces. He said, "You can still share three pieces and go on sharing forever." It might be fun to see how much more pizza is left after each student consumes the first four diminishing pieces.

All of the students got a nice feel for limits as n approaches infinity. There is often that perceptive student who says just cut it from the vertex to a base divided in thirds (1/3) since the remaining piece is going to get real small real fast.

Quadrilaterals Flying High

Geometry—Teacher Notes

Overview	
Students investigate the properties of quadrilaterals.	**Prerequisite Understandings** • Basic vocabulary of quadrilaterals. • Congruency of triangles.

Curriculum Content	
CCSSM Content Standards	G.CO.10L. Prove theorems about triangles. Theorems include: measures of interior angles of a triangle sum to 180°; base angles of isosceles triangles are congruent; the segment joining midpoints of two sides of a triangle is parallel to the third side and half the length; the medians of a triangle meet at a point.
CCSSM Mathematical Practices	3. **Construct viable arguments and critique the reasoning of others**: Students make conjectures and justify their conclusions on an investigation of quadrilaterals. 4. **Model with mathematics**: Students will create scale drawings and use measures and transformations to justify properties of quadrilaterals and their diagonals.

Task	
Supplies • Graphing paper/Patty paper • Rulers (or tape measures) • Compasses • Tables (or large butcher block paper) for the original size kite	**Core Activity** Students explore kites by making a full-sized model (on tables using dry erase markers or butcher paper) and then making a scale model on with centimeter paper. Conjectures are made and checked using measurements and transformations.
Launch Students will make a scale drawing of a rectangle and be guided through the methods of creating and justifying conjectures.	**Extension(s)** This activity opens up room for exploration into making large models of many geometric shapes and discovering the properties of geometric figures.

Quadrilaterals Flying High

Launch

A landscape company is given the job of creating a square garden with 20-foot sides (with diagonals!) for the new Math Museum. They must also design a plaque to discuss the mathematics of the square and its diagonals. Your task is to help them out!

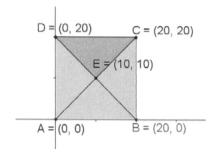

1. Make a drawing of the square and the diagonals with a scale of 1 cm = 1 ft.

2. Make a table of the measures of all the segments.

3. Make a table of the measures of all the angles.

4. Make a list of all the triangles congruent to $\triangle DCE$. How can you prove they are congruent using the triangle congruency conjectures?

5. How can you prove the triangles are congruent using patty paper and transformations?

6. Here is a list of some conjectures that can be included on the plaque. Explain how you can tell which ones are true and which ones are false.

    - The diagonals are perpendicular bisectors of each other.
    - All the sides of the square are parallel.
    - $\triangle AED \cong \triangle DCB$
    - $\overline{AE} \cong \overline{BE}$
    - A diagonal will create two (2) right isosceles triangles.
    - Diagonals create complimentary angles.

7. Make a list of the properties of this drawing, and post them on the board. As you see new ones **that you believe are true**, add them to your list; and if you see any **that you think are false**, mark them with a question mark for the class to discuss.

8. Design the plaque to describe the garden.

Quadrilaterals Flying High

Activity

Billie Jean decided to really make a great scale model of a kite that would just fit on her math table at school. Her table is 5' by 30". Your job is to do the same thing. Make a scale drawing of a kite that would fit on the table. Make sure your kite is a **Geometric Kite**.

1. Make a scale drawing of the table and the kite on another sheet of paper. Use a coordinate system for the drawing.

2. Copy the drawing of the kite and measure all the angles, slopes, and lengths. You can make a model on your desk with a dry erase marker to use in measuring.

3. Describe how the angles and sides of the kite compare. Write your discovery as a property. Ex: In a kite, the pairs of adjacent sides are congruent.

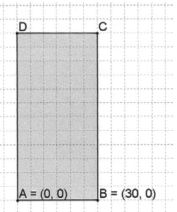

4. Now make another drawing of the kite with the diagonals drawn in.

5. Measure all of the angles' slopes and all of the lengths of the new segments created.

6. Make a list of all the observations you can make based on this drawing.

7. What triangles are congruent? Explain how you can tell using transformations, conjectures, and the fact that corresponding parts of congruent triangles are congruent.

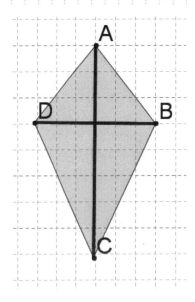

Quadrilaterals Flying High

Results from the Classroom

This problem was presented right after students had finished constructions and had started to do triangle properties. They had not done any coordinate geometry, and it is interesting to see how well they did with a scale drawing, how well they put a coordinate system on the graph, and how exact they tried to be.

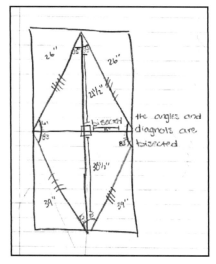

The drawing at the left was representative of many of the ones turned in. Students were able make a sketch, but the drawings were not to scale or accurate. It seems that our lack of emphasis on measurements (from using a compass and straightedge) didn't prepare them for scale drawings at all. By not making a true scale drawing, the shape on the left became a rhombus and the conclusions actually lead to erroneous ideas.

The next drawing (below) was by a student who asked for an architect's scale. This better result suggests how the appropriate use of tools can impact results.

The conclusions here were typical of many students.

One student looked at most of the project globally. His drawings were sketches with accurate measures, but his conclusions were also global and descriptive such as the "top" two and "bottom" two triangles are congruent.

His global conclusions helped push the class toward generalizations more than the specific triangles like those to the right.

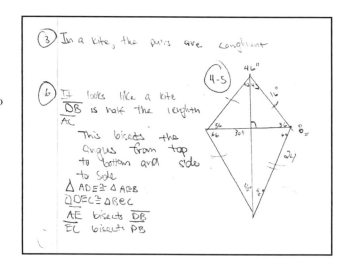

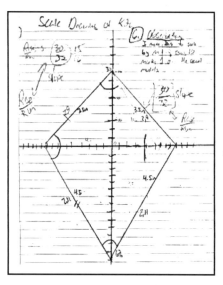

The last example (at left) shows the best drawing at using a coordinate system with a clearer picture of the student's ideas related to slope and its relationship to geometric drawings.

This Great Task could be adapted to fit the unit on constructions, the unit on congruent triangles and quadrilaterals, the unit on coordinate geometry, or be kept as a summative task for all of those units.

Classy Carnival

Statistics and Probability—Teacher Notes

Overview	
Students will create a carnival game, predict the expected value, and then simulate the experiment.	**Prerequisite Understandings** • How to simulate a problem.

Curriculum Content	
CCSSM Content Standards	S.MD.2. (+) Calculate the expected value of a random variable; interpret it as the mean of the probability distribution.
CCSSM Mathematical Practices	3. **Construct viable arguments and critique the reasoning of others:** Students will create a game and then justify their design. 4. **Model with mathematics:** Students will create a simulation to test their reasoning.

Task	
Supplies • Spinners or random number generators	**Core Activity** Use the task as a group springboard to creatively analyze carnival type games.
Launch The launch allows the class to look at expected values and discuss ways to compute them.	**Extension(s)** The next level is to evaluate skill-level games. When does a person's skill help them to beat a carnival game, and how could you factor that into the computation of the expected value of a game?

Additional Resources

Appendices A & B contain instructions for **Generating Random Integers** operations on both the TI-Nspire™ handhelds and the TI-84 Plus graphing calculator.

Classy Carnival

Launch

Sweet Tam the Stats Teacher likes to use a spinner to determine the number of problems that the students get each night for homework. They can end up with 0, 5, 8, 10, or 16 problems each night.

1. What is the probability of landing on any of the sectors?

2. How can you justify your answer by angles and by areas?

3. Using this spinner, explain how you can figure out the average number of problems you can expect each night from Sweet Tam.

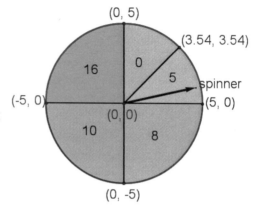

Classy Carnival

Activity

The Math Club is looking to raise a couple of thousand dollars for a trip to the new MoMath museum in New York City! Check it out at Momath.org

You have been selected to design a game for your school's Classy Carnival to help raise funds. Here are the guidelines:

1. Your game must be based on probability, not skill!

2. You must post the following information about your game:
   a. The rules to play the game.
   b. Chances of winning each prize.
   c. Cost to play the game.
   d. Values for each prize.

3. You will also have a secret one-page document that will show:
   a. Expected profit if 100 people play the game.
   b. Calculations for how you figured the profit.

Write up a proposal for the game based on the rules, and show how you came up with the calculations for the posted information.

Classy Carnival

Results from the Classroom

This task is a springboard for developing the concept of expected value. Students are intuitively set to thinking about what the chances are of achieving all outcomes and then using the payoff for all the outcomes for compute a net return from the game. The creativity is also entertaining and fun for students.

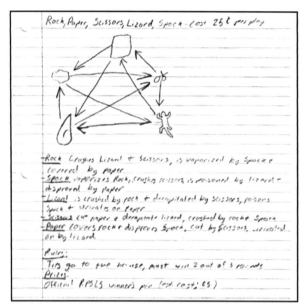

This first game is a variation on Rock-Paper-Scissors but with a twist! Students have multiple ways of winning and losing based on the game and then the house has the advantages when there is a tie. The students decided they could earn a marginal $24.78 out of 100 games. The answer clearly needs some explanation!

A second game had the students running a random robot around the floor for 10 seconds while blindfolded. This was a fun twist on a random spinner and the probability was based on putting numbers on the floor of returns. The students predicted a $150 profit on 100 $5 tries - certainly a bit optimistic in terms of price.

Another fun idea was based on a spinner with a reported 1.73% chance of winning a $10,000 car for a $3 ticket. The students' belief that this would create a $300 profit on 100 plays generated some interesting comments from other students and resulted in a much better understanding of expected returns for the originators of the game.

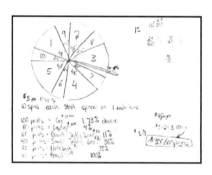

Another group created a race car game with a spinner for allowing students to move on a board against other competitors. The return is based on having lots of players who love to compete against each other. Only the winner collects more than he paid.

As a follow-up problem, have students find the expected return on a horse race game. The house gets horse #7 while all the other players get the horses numbered from 2 to 12. Two dice are rolled and the horse with the winning sum moves ahead on a 5-square race. Since there are more ways to make a 7 with two dice than any other number, the #7 horse has a built-in chance to move 1/6 of the time while the poor people who get horses 2 and 12 only have a 1/36 chance to move a space on each toss of the die.

Will people find these rigged games worth playing? Sure, one only has to look at the state lotteries.

Coaches' Dilemma

Statistics and Probability—Teacher Notes

Overview	
Students will develop a simulation to understand a conditional probability problem and then apply the mathematics to determine the probabilities.	**Prerequisite Understandings** • Ability to calculate basic and conditional probabilities.

Curriculum Content	
CCSSM Content Standards	S.CP.3. Understand the conditional probability of A given B as P(A and B)/P(B), and interpret independence of A and B as saying that the conditional probability of A given B is the same as the probability of A, and the conditional probability of B given A is the same as the probability of B.
CCSSM Mathematical Practices	4. **Model with mathematics:** Students use simulations to explore branching choices in probability.

Task	
Supplies • Random number generators (calculators, cards, dice, spinners, etc.) for simulation	**Core Activity** The core task is a revision of the Monty Hall Problem. Students are allowed to explore and to use the probabilities in a different situation that they might find more familiar. The problem involves conditional probabilities, but it can be worked without the formula and then the formulas can be introduced to students following the activity.
Launch Students should work problems which require the use of conditional probabilities, recognizing that many can be done without the use of the formula.	**Extension(s)** Students could work more problems involving conditional probabilities and look for real-world applications of this concept.

Additional Resources

Appendices A & B contain instructions for **Generating Random Integers** on both the TI-Nspire™ handhelds and the TI-84 Plus graphing calculators.

Coaches' Dilemma

Launch

The probability of two events occurring in a row is multiplicative. That is, if the probability that it will rain on Tuesday is 75% and the probability of rain on Wednesday is 60%, then the probability that it will rain on Tuesday and Wednesday is 75% of 60% or $0.75 \cdot 0.60 = 0.45$ or 45%.

A jar contains quarters and pennies. Two coins are chosen without replacement. The probability of selecting a quarter and then a penny is 0.34, and the probability of selecting a quarter on the first draw is 0.47. What is the probability of selecting a penny on the second draw, given that the coin first drawn was a quarter?

$$P(\text{Penny} \mid \text{Quarter}) = \frac{P(\text{Quarter and then a Penny})}{P(\text{Quarter})} = \frac{0.34}{0.47} = 0.72 = 72\%$$

1. The probability that it is Monday **and** that a student is absent is 0.04. Since there are 5 school days in a week, the probability that it is Monday is 0.2. What is the probability that a student is absent given that today is Monday?

2. At Modern Middle School, the probability that a student takes Chorus **and** French is 0.067. The probability that a student takes Chorus is 0.88. What is the probability that a student takes French given that the student is taking Chorus?

3. At a middle school, 22% of all students play soccer **and** basketball, and 36% of all students play basketball. What is the probability that a student plays soccer given that the student plays basketball?

4. In the United States, 56% of all children get an allowance, and 30% of all children get an allowance **and** do household chores. What is the probability that a child does household chores given that the child gets an allowance?

5. In Europe, 92% of all households have a television. 68% of all households have a television **and** a DVD player. What is the probability that a household has a DVD player given that it has a television?

6. A college professor gave her class two tests. 25% of the class passed both tests, and 42% of the class passed the first test. What percent of those who passed the first test also passed the second test?

Coaches' Dilemma

Activity

The coaches are setting up the schedule for the spring basketball tournament. This year, they want to have some fun and try an unusual technique since there was not enough gym space to play a round robin tournament. While the coaches were meeting with the tournament director, they came up with a plan to use a probability activity.

Your team has a choice of three opponents. Team A is undefeated, and Team B has a 14-2 record, while Team C has yet to win a game. Since your team has a 10-4 record, your coach knows your best chance of being in the winner's bracket is if you can play Team C.

They decide to put the names of each team in a different bag labeled 1, 2, or 3. Your coach picks a bag. Then, to make it more interesting, the tournament director opens one of the two bags remaining to show where one of the best teams was located. Now, your coach can stay with the original choice or can switch to the other unopened bag.

1. Which choice will yield the best chance of playing the team with the losing record, Team C, first? What is the probability of winning with 'the stay' and with 'the switch' strategies?

2. Devise a different strategy for seeding 4 teams. Describe how each of the 4 teams could use conditional probability to evaluate their chances of winning.

3. Devise and briefly discuss a method to give an advantage to teams with the least probability of winning.

Coaches' Dilemma

Results from the Classroom

This is a strange way to set up a tournament, yet students didn't question the logic of letting one teacher choose to switch based on some information that would clearly impact the outcome of the tournament. They all seemed to easily accept the fact about gym space, choosing teams from a bag, and getting to change your opponent.

This task was not difficult for most students. Therefore, it is advisable to extend the task to evaluate other strategies of seeding 4 teams in a tournament and how those strategies would impact a team's chances at success.

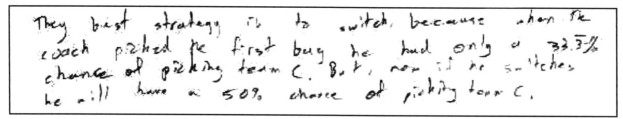

The student above didn't need a great deal of discussion to explain his solution.

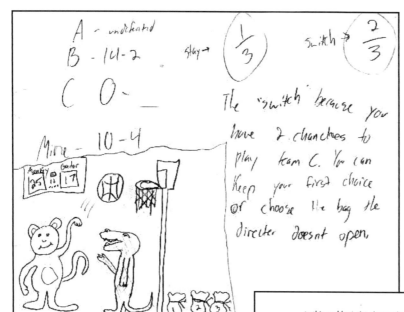

The student to the left added a little fun with a picture. His incorrect probabilities did not seem to counteract his belief that two chances are better than one.

Another student focused on the risk but did not really look at the probabilities. He was sure that the team revealed by the athletic director might include the best team which is an incorrect interpretation of the problem.

In this set of students, the conditional probabilities did not come up. It would be a good opportunity to introduce them and lead into more real world applications.

> In this problem, the chance that you picked C will always be ½. No matter what bag you picked, one of the other two bags will always be revealed to be one of the better teams. That eliminates one of the total options. This bag will never be C. This leaves us with two bags. Theoretical probability suggests that the bag you chose from the beginning will have a chance of ½ of being the C bag. If you switch, the probability remains the same. The real problem is the risk. The athletic director will reveal one bag. If that bag happens to be the best team, then the risk of playing the best team is completely eliminated. If the bag is the second best team, then there is a greater risk of playing the best team, which is the most undesirable outcome. All in all, you will always have the same chance of picking C. If I were in that situation, I would always stay with my choice, because it really wouldn't matter.

Statistics and Probability

How Fast Does it Fall?

Statistics and Probability—Teacher Notes

Overview	
Students will design an experiment to collect data on the height and time involved in dropping a parachute.	**Prerequisite Understandings** • Data collection and organization.

Curriculum Content	
CCSSM Content Standards	S.ID.6. Represent data on two quantitative variables on a scatter pot, and describe how the variables are related. a. Fit a function to the data; use functions fitted to data to solve problems in the context of the data. *Use given functions or choose a function suggested by the context. Emphasize linear, quadratic, and exponential models.* b. Informally assess the fit of a function by plotting and analyzing residuals. c. Fit a linear function for a scatter plot that suggests a linear association.
CCSSM Mathematical Practices	4. **Model with mathematic:** Students create a table and find a line or curve to fit the data. 7. **Look for and make use of structure:** Students assess the closeness of fit.

Task

Supplies	Core Activity
• Cellophane or tissue paper • String • Masking tape • Paper clips • Scissors • Rulers or tape measures • Hole punch • Stopwatch • Graphing calculators	Working in teams of three or four and combining teams to aggregate data, students use graphing calculators to enter the data, create a scatter plot, and find the line of best fit.
Launch	**Extension(s)**
The launch activity reinforces the basic idea that a perfect linear function is still a 'good' fit if adding some additional points still results in a 'reasonable' relationship.	Aggregate data from the whole class and repeat the analysis. Compare students' initial results with those of the larger data sets. How do the two differ? Why would you expect these differences? Which model is a better predictor? Why?

Additional Resources

Appendices A & B contain instructions for **Entering Data, Plotting Data,** and **Calculating Regression Equations** on both the TI-Nspire™ handhelds and the TI-84 Plus graphing calculator.

How Fast Does it Fall?

Launch

1. Consider this data set, and find a function that describes the relationship.

X	-1	1	2	2.5	3	4	5		
Y	1	5	7	8	9	11	13		

2. If I add some additional points, would the function still be a good fit? Why or Why not?

3. In the table below, could x represent how many hours you worked for your parents and y represent how much you earned? Explain your thinking.

X	-1	1	2	2.5	3	4	5	3	5
Y	1	5	7	8	9	11	13	10	13.5

4. What does it mean to create a line or curve of best fit?

5. How do we know when we have a good fit?

How Fast Does it Fall?

Activity

Students at Gatineau High School are planning to enter a contest to see who can parachute an egg from the highest distance without the egg being damaged. In preparation for the event, the group decides to collect some time and distance data using a parachute constructed with tissue paper.

Part A. Construct parachutes for trials.

1. Cut a 7 cm square from tissue paper.
2. Place a small piece of tape on each corner and punch a hole through the taped area.
3. Cut four pieces of string of 42 centimeter length.
4. Tie a piece of the string to each of the four corners.
5. Tie the hanging ends of the string together at the bottom.
6. Add four small paper clips for the load.

Part B. The Experiment

Before you begin the experiment, consider the following:

1. Discuss the design of your experiment with your team members. Describe how you will carry out the experiment.
2. What are some potential sources of variability? How will you reduce these possible effects?
3. Drop your parachute from a minimum of ten different heights (record distances in centimeters), and record the time (in seconds) it takes it to reach the ground. Use the table on the next page to record the data from your trials. You will share your data with two other groups so you have a larger data set for the analysis.

Part C. The Analysis

1. Construct a scatter plot of distance versus time.
2. Discuss the data patterns with your team members. Does a linear model appear to be appropriate for describing this pattern? Explain.
3. Find a line of best fit for the data. Graph the line on the scatter plot. Sketch that graph, and write the function for the line of best fit.
4. Is a linear model a good fit? Why or why not?
5. What does the slope of your function tell you about the relationship between the height of the parachute drop and time? What does the y-intercept tell you?

Trial	Height (in centimeters)	Time (seconds)	Height (in centimeters)	Time (seconds)	Height (in centimeters)	Time (seconds)
1						
2						
3						
4						
5						
6						
7						
8						
9						
10						

How Fast Does it Fall?

Results from the Classroom

The introductory activity was a good way to review finding the equation of a line as well as some general ideas about line of best fit.

The first data set presented results in a linear function with an equation of $y = 2x + 3$. It is a good idea for students to plot these points and then add the additional points (3, 10) and (5, 13.5) to the plot.

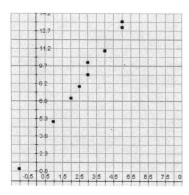

Students will note that the "additional points are approximately the same as the one in the first set" or "the two new points are very close to the line formed by the original set, so it is still a pretty good fit". [The line of best fit for the data set with the two additional points is $y = 2.05x + 3.04$. It is isn't expected that students would find the line of best fit, though *Illuminations* on the NCTM website has an applet (Line of Best Fit) that allows this to be easily done.]

Students describe their experimental design for the activity while the teacher monitors their work. Having the students describe their design provides a time for reflection as well as an opportunity to point out potential sources of variability (such as different students dropping the parachute in different ways; errors in measurement; breezes or drafts; location; etc.).

Students' work varies. One group shared the following data {(147, 1.9), (148, 2.0), (149, 2.2), (150, 2.3), (151, 2.6), (152, 2.7), (153, 2.8), (154, 3.0), (155, 3.1), and (156, 3.4)}. Their graph and resulting line of best fit (completed using Excel) appear below:

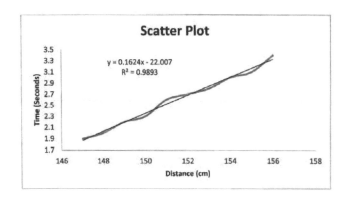

Students described their pattern as "a linear model where a trend line fits well with the data points". They also offer "Time or $y = 0.1624$ (distance or x) $- 22.007$ is a good approximation for how the data points are distributed." Students may offer their R^2 value as a measure of 'goodness of fit'. In the student example, R^2 is 0.9893. Values can range from 0 to 1. A value close to 1 indicates that the model accounts for a greater proportion of variance.

Student discussion of slope should get at the idea that slope "gives the rate of change of time with respect to the rate of change of distance or height." It is important for students to also know that the *y*-intercept is the value for which x is 0 "which is not logical to think of dropping the parachute from a height of 0 – so the domain has to be greater than 0".

Searching Similarities

Statistics and Probability—Teacher Notes

Overview	
Students will explore data categorization to calculate frequency and select appropriate graphing strategies.	**Prerequisite Understandings** • Ability to draw frequency tables, bar graphs, line plots and pie graphs. • Ability to calculate probability and odds.

Curriculum Content	
CCSSM Content Standards	S.ID.1. Represent data with plots on the real number line (dot plots, histograms, and box plots.) S.ID.2. Use statistics appropriate to the shape of the data distribution to compare center (median, mean) and spread (interquartile range, standard deviation) of two or more different data sets. S.ID.5. Summarize categorical data for two categories in two-way frequency tables. Interpret relative frequencies in the context of the data (including joint, marginal, and conditional relative frequencies). Recognize possible associations and trends in the data.
CCSSM Mathematical Practices	4. **Model with mathematics:** Students will categorize people or animals into sets with categories. 7. **Look for and make use of structure:** Students will use rubrics or rules for categorization to help determine the structure.

Task

Supplies

- Copies of pages from "search" books. An example would be Martin Handford's *Where's Waldo?* series.
- Poster board for graphs
- Color markers
- Rulers and compasses
- Graph paper
- Calculators

Core Activity

Students search for similarities to categorize data from concrete examples. They then calculate frequencies and create graphs of the data.

Launch

The activity introduction will review frequency distributions, bar graphs, and pie graphs and how to calculate probability vs. odds.

Extension(s)

Choose a second way to categorize the people. Make comparisons using other forms of graphs. Find probabilities of compound events.

Searching Similarities

Launch

1. List at least five different types of graphs.

2. Describe what types of data can be displayed in each graph. What is missing in each of the graphs listed above? Why would that be important?

3. Look at the newspaper and find a graphic display. Name the type of graph, and describe everything you can learn about the information presented in it. Is there any information missing that might be helpful? How does that contribute to the opinion being presented by the author?

4. Often some types of graphs are selected so that information can purposely not be reported. Describe a situation where that might be advantageous.

5. Look at the newspaper for situations where probability is involved. Discuss why it is necessary and how it might have been calculated.

6. Discuss attributes that can be can be graphed and how a graph could be created to identify two attributes at the same time. Which graph type would be the most informative?

Activity Instructions

Students will each choose a picture from a book which contains searches. An example would be Martin Handford's *Where's Waldo?* series. The students will decide how to categorize the people in the picture, or, if it is another type of book, the animals or items. Students must define how they have categorized their people so that each person fits in only one group. They should ask you to check before they start counting people.

Be sure students have access to examples of each type of graph. Hand out the activity sheets and discuss the problem—note the step-by-step questions to be filled in as the student progresses. Students may work with a partner to tabulate the information from Waldo page.

Build in discussion time for the class to explain their graphs and conclusions.

Searching Similarities

Activity

1. Choose one of the pictures from the book and find the indicated item. Write a description of the method you used to find the selected item. How did you find it? Can you think of another way?

2. Working with a partner, decide how to categorize the people or items in the picture. You can set up a comparison such as clothing color, headgear, objects held, etc. depending on the picture you choose.

    - Describe how you are categorizing your people, and why you choose this way to divide them.
    - Make a table with a frequency distribution to show how many of each person fit into each category.

Note: The headings for the columns below are suggestions only. Use your own as needed.

Category/Criteria/Grouping	Additional Category/Tally	Frequency

3. Calculate the angles, and draw a pie graph documenting your calculations.

4. Construct a histogram and a line plot showing totals of the categories.

5. Calculate probability that the item fits in each of your categories. Explain how you calculated the probability.

6. Calculate the odds that the item fits in each of your categories. Explain how you calculated the odds.

7. Look at the data again. Is there something that might categorize the people in two different ways? For example, could you look at age and actions or maybe color of clothing and head gear? Would all of the data points have both? Create the type of chart you would need to summarize this data.

8. Find the probabilities and at least two conditional probabilities associated with your data.

9. Describe a situation where you believe conditional probabilities could be helpful.

Searching Similarities

Results from the Classroom

The students had a variety of ways to find the missing person or item.

divided picture & carefully looked at each part + looked for stripes

I scan sections of the page from left to right.

I looked for his shirt pattern.

Color was an attribute that was often used in setting up the first graph. When asked to select a second attribute, the students again looked for attributes that were simple to categorize. Hair color or actions were often used. The students rarely had trouble when dealing with one type of data, but they really struggled with putting two types of data into a chart.

Clothing Color	Tally	Frequency
Yellow	|||| |||| |||| |||| |||| |||| |||| |||| ||||	50
Blue	|||| |||| |||| |||| |||| |||| |||| |||| |||| |||| |||| |||| |||| |||| |||| |||| ||	87
Red	|||| |||| |||| |||| |||| |||| |||| |||| |||| |||| |||	58
Green	|||| |||| |||| |||| |||| |||| |||| |||| |||| |||| ||||	54
Brown	|||| |||| |||| ||||	19
Gray	|||| |||| ||||	15
		282

Many students needed help creating structure from the data. The probabilities varied, and collecting data from the pictures was helpful to the students by providing a concrete model. Statistics is of more interest when they can become involved. Students struggled to find and describe real world examples of their own though.

	Eating	Serving	Cooking
Blue	20	31	5
Red	12	5	4
Yellow	10	3	1
Green	16	8	4
Black	33	7	2

The $P(R/E) = \frac{12}{21}$

YouTube™ Views

Statistics and Probability—Teacher Notes

Overview	
Students will use scatter plots to draw conclusions from given data and to evaluate conjectures made on the basis of the data.	**Prerequisite Understandings** • Creating and interpreting scatter plots. • Using and interpreting simple descriptive statistics.

Curriculum Content	
CCSSM Content Standards	S.ID.5. Summarize categorical data for two categories in two-way frequency tables. Interpret relative frequencies in the context of the data (including joint, marginal, and conditional relative frequencies). Recognize possible associations and trends in the data. S.ID.6. Represent data on two quantitative variables on a scatter plot, and describe how the variables are related.
CCSSM Mathematical Practices	2. **Reason abstractly and quantitatively:** Students have to take an open set of data and analyze the data by creating a way to evaluate the effectiveness of videos. 3. **Construct viable arguments and critique the reasoning of others:** Students defend their thinking to the class.

Task	
Supplies • Graph paper or statistical graphing technology (spreadsheet or calculator)	**Core Activity** Students use spreadsheets as a good way to organize the data, calculate ratios, and create statistical graphs to help justify their conclusions.
Launch The Launch activity takes students through the process of interpreting a scatter plot and helps them to critically read the scales and the data.	**Extension(s)** Gather real data on various videos and establish measures or criteria for a realistic measure of a successful video. Students could create a video about creating and analyzing scatter plots.

Additional Resources

Appendices A & B contain instructions for **Entering Data** and **Plotting Data** on both the TI-Nspire™ handhelds and the TI-84 Plus graphing calculator.

YouTube™ Views

Launch

The graph below shows the amount of money earned compared to the amount of money saved for 11 high school students during the month of July. An estimated line of best fit and its equation are shown.

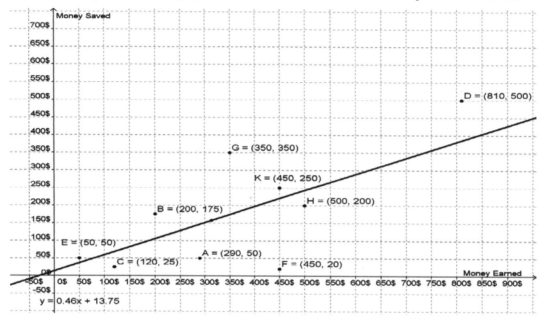

July Money Math 1

Based on this information, answer the following questions to demonstrate your understanding of scatter plots and lines of best fit.

1. Describe what point G (Grace) tells you about Grace during the month of July?

2. If Hewey is represented by point H, what are Hewey's earnings, savings, and spending for July?

3. What can you say about the people above the line? Do they save more or less than the average student?

4. What is the average amount these 11 students earned, saved, and spent?

5. Is the average amount spent the same as the difference between the average amount earned and the average amount spent?

6. Using this data, what can you conclude about the % of wages that these students saved?

7. The equation for the estimated line of best fit is $y = .46x + 13.75$. What do x and y represent and what does the .46 mean in terms of earnings and savings?

8. Which of the graphs on the right shows a reasonable correlation?

9. If the graph above had a negative correlation, what would it say about the saving habits of the students?

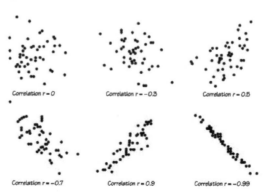

Statistics and Probability

YouTube™ Views

Activity

Five friends are competing to produce the most popular video for YouTube on how to create scatter plots. To help decide who won, they gathered the data about their videos shown in the table below.

Student	Number of Views	Number of Days Posted	Length of video
Molly	932	10	3 min 20 sec
Frankie	1075	8	3 min 0 sec
Kendra	866	17	4 min 0 sec
Helga	466	30	7 min 12 sec
Luke	718	25	5 min 10 sec

1. Evaluate the friends' success using mathematics and statistics. Decide who won this competition, and justify your decision using tables, ratios, equations, and graphs.

2. Molly conjectured that the length of videos was related to the number of views. Does the data support the conclusion that length matters? Explain your reasoning.

3. Debate your conclusion with the rest of the class.

4. If you were really trying to test your conclusions, what kinds of additional data should you examine in order to convince someone that you had a reasonable rating system to see what YouTube videos are the most popular with viewers? How could you categorize videos so that the ratings would be useful within a specific category (examples: educational videos or music videos)?

Statistics and Probability

YouTube™ Views

Results from the Classroom

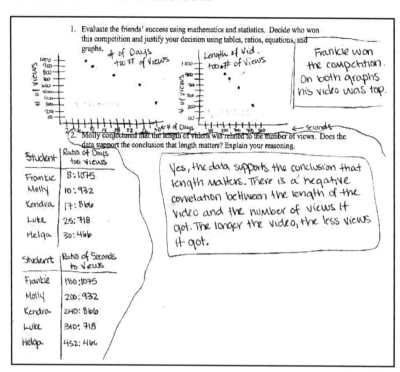

The title of the student work could have been 'you get what you teach.'

This task was given to students in the traditional curriculum after 3 years of high school mathematics including Algebra 2. What is noteworthy about this analysis is the students' level of work does not reflect the common core standards listed in the teacher notes.

Why the division? Clearly our current curriculum does not lend itself to a high level of expectation for categorical data, and the task itself does not demand that students look at that type of analysis.

We get some rather nice direct conclusions made from the application of descriptive statistics from students.

The task should be piloted with students in a Statistics class to see if it would work with students who understood that the data are minimal and leads students to discuss the need for gathering more information before making too many conclusions.

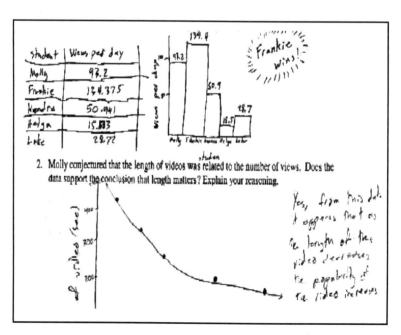

Statistics and Probability

Appendix A:
TI-Nspire™ – Plotting and Analyzing Data

Entering data into a spreadsheet

Press [⌂on] and select New Document to start a new document.

Choose **Add Lists & Spreadsheet.**

Note: To add a Lists & Spreadsheet page to an existing document, press [ctrl] [doc▾] and choose Add Lists & Spreadsheet. Alternatively, press [⌂on] and select [▦].

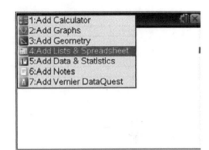

In column A, press ▲ on the touchpad to move to the cell at the top of column A. Be sure to move to the top of the column. Alternatively, click in the cell at the top of column A.

Type the list name **folds** next to the letter A, and press [enter].

Move to the cell at the top of column B, type the list name **layers**, and press [enter].

Note: The data entered below are from the "paper folding" activity that many teachers use to introduce exponential functions.

Enter data, as shown, into the two lists.

Place the cursor in cell A1 for column A or in cell B1 for column B. Enter the first number.

Note: Pressing [enter] or ▼ will move the cursor to the next cell in the column.

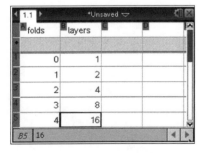

Calculating one-variable statistics

When you have statistical data stored in a list, you can display the one-variable statistics for the set of data.

Move the cursor to cell B1.

Press **MENU** > **Statistics** > **Stat Calcualtions** > **One-Variable Statistics**.

Press **OK** to analyze one list (layers).

The X1 List is stored in column B. Alternatively, use the right arrow to select the name of the list – layers. Press **a**.

Use the arrow keys as needed to view the displayed statistics.

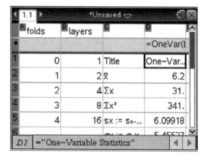

Plotting data

When you have statistical data stored in lists, you can display the data you have collected in a scatter plot, connected scatter plot, dot plot, histogram, box plot, normal probability plot, or summary plot.

After entering data in a Lists & Spreadsheet page, press [ctrl] [doc▼] to add a new page to the document. Select **Add Data & Statistics**.

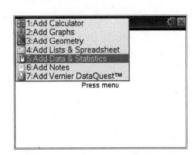

Note: When the Data & Statistics page is added, the list name that appears next to the word "Caption" is the first list name, alphabetically, in the spreadsheet. However, if there are any categorical data lists, the first categorical list name, alphabetically, will be displayed when the Data & Statistics page is added.

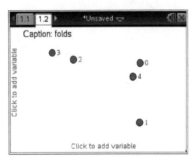

To choose the variable for the horizontal axis, move the cursor to the "Click to add variable" message at the bottom of the screen. Press [👆] to display the variables.

Alternatively, after adding the Data & Statistics page, press [tab] to display the variables available for the horizontal axis.

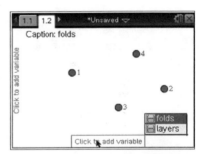

Select the variable folds.

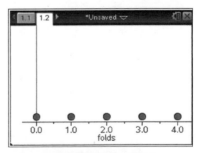

Move the cursor to the middle of the left side of the screen. When "Click or Enter to add variable" appears, press [👆] to display the variables. Select the variable layers.

Alternatively, after adding the horizontal axis variable, press [tab] to display the variable choices for the vertical axis.

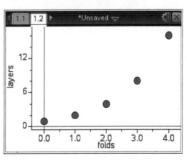

Appendix A   175   TI-Nspire™ Plotting and Analyzing Data

Calculating a Regression Equation

To calculate and display a regression equation on a Data & Statistics page, press **MENU > Analyze > Regression > Show Exponential**.

Note: Use the type of regression appropriate for your data.

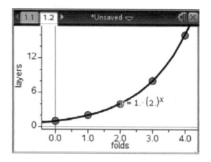

If you need to calculate a regression equation and display the statistics, do so on a Lists & Spreadsheet page. Return to the Lists & Spreadsheet page using `ctrl` ◄.

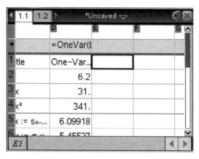

Move the cursor to the first empty column. **Press MENU > Statistics > Stat Calculations > Exponential Regression**.

Set up the dialog box with the X List and Y List. Arrow down to ensure that the 1ˢᵗ Result Column will not replace existing data. Press **OK**.

Use the arrow keys as needed to view the displayed statistics.

Note: Alternatively, you could calculate the regression equation on a Calculator page. However, the calculation will not be dynamic in the event that you edit data in the lists. The regression equation calculation on the Lists & Spreadsheet page is dynamic.

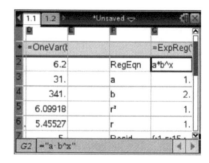

Appendix A

Generating random integers

On a Calculator page, press **MENU > Probability > Random > Seed** and enter a unique starting number for the seed value. Press [enter].

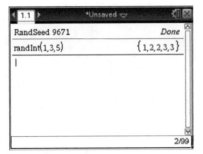

Press **MENU > Probability > Random > Integer**.

Enter the lower bound, upper bound, and if desired, how many random numbers you want generated. Press [enter].

The command shown in the screen capture generates five random numbers between 1 and 3.

Graphing parametric equations

Insert a Graphs page. Press [ctrl] [doc▼] and choose **Add Graphs**.

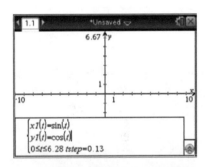

Press **MENU > Graph Entry/Edit > Parametric**.

Enter x(t) and y(t).

Press [enter] to graph.

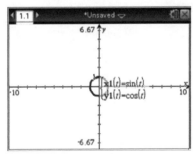

Recursion using previous calculation

Insert a Calculator page. Press [ctrl] [doc▾] and choose **Add Calculator**.

Press **4×2** and [enter].

Press **×2** and [enter].

Press [enter].

Continue to press [enter].

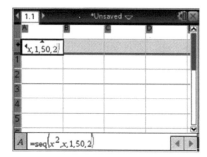

Generating a sequence in a list

Insert a Lists & Spreadsheet page. Press [ctrl] [doc▾] and choose **Add Lists & Spreadsheet**.

Move to the formula cell for column A.

The syntax for the sequence command is the following:

seq(Expr,Var,Low,High[,Step])

To generate a list of the squares of the odd numbers from 1 to 50, use the command:

seq(x²,x,1,50,2)

Press [enter].

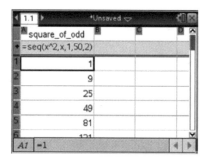

Note: The step size is 1 by default. The brackets around Step in the syntax for the command indicate that you only need to include the step size when it is not equal to 1.

Appendix A　　　　　　　　　　178　　　　　　　　TI-Nspire™ Plotting and Analyzing Data

Graphing a sequence

Follow the directions above for Generating a Sequence in a List. Insert a Graphs page.

Press **Menu > Graph Entry/Edit > Sequence > Sequence**.

The following example generates a sequence of the squares of the odd integers greater than 0.

Press **Menu > Window / Zoom > Zoom-Fit** for the viewing window shown in the screen captures or use **Window Settings** to set up a custom viewing window.

Press **Menu > Table > Split-screen Table** to view (or remove) the table.

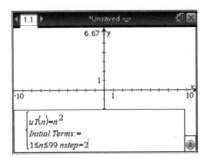 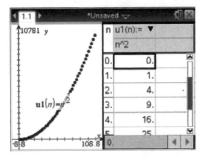 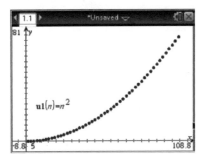

The following example generates the Fibonacci sequence, where u(n)=u(n-2)+u(n-1).

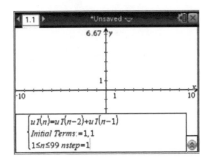 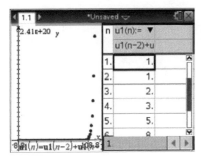 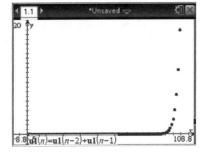

Appendix B:
TI-84 Plus – Plotting and Analyzing Data

Entering data into lists

Create L_1 and L_2 from the given information: (2, 6), (4, 5), (5, 3), (6, 1), (8, 0)

Press

This command takes you to the statistical editor. If your editor does not currently show L1, L2, and L3, then press [STAT] [5] [ENTER] to set up the editor. Press [STAT] [ENTER] again to re-enter the editor.

(if L1 already contains data)

(if L2 already contains data)

[◄] returns to L1.

Enter data in L1 and L2 using the numeric keys, [ENTER], and the arrow keys to move between lists.

[2nd] [MODE] exits the data editor and displays whatever is currently on your Home screen.

Calculating one-variable statistics

When you have statistical data stored in a list, you can display the one-variable statistics for the set of data.

 selects 1-Var Stats.

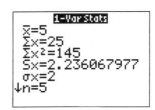

(if L1 contains the data and the frequency of the list is 1.)

Press to view additional stats.

Plotting data

When you have statistical data stored in lists, you can display the data you have collected in a scatter plot, xyLine, histogram, box plot, or normal probability plot.

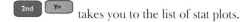 takes you to the list of stat plots.

 turns all plots off if any plots are on.

 selects Plot 1.

ENTER turns Plot1 on.

Use the left/right arrows and ENTER to select the type of plot. The first one is a scatter plot.

 enters L1 as the Xlist.

 enters L2 as the Ylist.

Use the left/right arrows and enter to select the plotting mark.

Press .

Use the up/down arrows to enter appropriate values for Xmin, Xmax, Ymin, Ymax.

Alternatively, use ZOOM > **ZoomStat** to set up the viewing window automatically based on the data in the lists.

Press GRAPH.

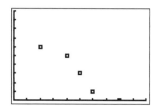

Note: If the calculator graphs another function besides the stat plot, press Y= to CLEAR any functions Y_1 through Y_{10} that are not empty and then GRAPH again.

Appendix B   182   TI-84 Plus Plotting and Analyzing Data

Calculating a Linear Regression Equation

Press [STAT] then use right arrow to move the cursor to CALC.

Select **LinReg(ax+b)**. Press [ENTER].

Tell the calculator which lists to use to create the linear regression equation and where to paste the equation once it is calculated. Press [2ND] [L1] [ENTER] for **Xlist** to be L1, [2ND] [L2] [ENTER] for **Ylist** for L2, and [ALPHA] [TRACE] [ENTER] to store the regression equation in Y1. Use the down arrow to highlight **Calculate**, then press [ENTER] to calculate the regression equation.

If you do not see r^2, then turn on your diagnostics by pressing [MODE], then [↓] until you reach **STATDIAGNOSTICS**. Press [▶] [ENTER] [CLEAR]. Then you can press [↑] on the Home screen until you highlight the original command, [ENTER] to paste the command to the current entry line, and [ENTER] to calculate the regression equation again with the diagnostics on.

The equation is pasted in the [Y=] screen. Press [GRAPH] to view the line of regression.

Appendix B   TI-84 Plus Plotting and Analyzing Data

Generating random integers

2nd **MODE** exits your current location and displays whatever is currently on your Home screen.

Press **MATH** > **PRB** > **randint(**.

Enter the lower bound, upper bound, and if desired, how many random numbers you want generated. Press **ENTER**.

The command shown in the screen capture at right generates five random numbers between 1 and 3.

Graphing parametric equations

Press **MODE** and highlight **PAR** for parametric graphing mode.

Press **Y=** to enter X1(t) and Y1(t).

Press **ZOOM** > **ZSquare** to graph with a square viewing window.

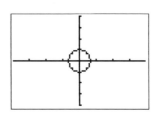

Appendix B   184   TI-84 Plus Plotting and Analyzing Data

Recursion using previous calculation

One the Home screen, press [4] [×] [2] and [ENTER].

Press [×] [2] and [ENTER].

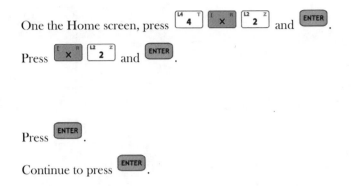

Press [ENTER].

Continue to press [ENTER].

Generating a sequence in a list

Open the statistical editor by pressing .

If needed, clear the data from L1.

Press ▲ to create a formula to generate data in L1.

Press  and choose **Ops > seq(**.

Set up the dialog box as shown to generate a list of the squares of the odd numbers from 1 to 50.

Press ENTER on Paste.

Press ENTER to calculate.

Graphing a sequence

Press .

Highlight **SEQ** and press ENTER for sequence graphing mode.

Highlight **DOT** and press ENTER so that the graph points will not be connected.

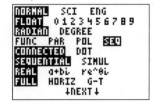

Press Y= to enter nMin and u(n). The following example generates a sequence of the squares of the integers greater than 0.

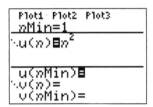 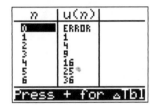 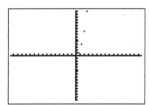

The following example generates the Fibonacci sequence, where u(n)=u(n-2)+u(n-1).

 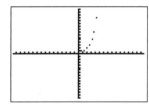

Appendix B   187   TI-84 Plus Plotting and Analyzing Data